Friction, Lubrication and Wear

Edited by
Mohammad Asaduzzaman Chowdhury

Published in London, United Kingdom

IntechOpen

Supporting open minds since 2005

Friction, Lubrication and Wear
http://dx.doi.org/10.5772/intechopen.77584
Edited by Mohammad Asaduzzaman Chowdhury

Contributors
Edgar Ernesto Vera Cardenas, Armando Irvin Martínez Pérez, Manuel Vite-Torres, Jose Luis Bernal Ponce, Karina Aleman Ayala, Marisa Moreno Rios, Sergey Fedorov, Walter Holweger, Hongbo Ju, Waldemar Tuszynski, Marek Kalbarczyk, Bartosz Kiser, Michal Pawel Michalak, Remigiusz Michalczewski, Jerzy Mydlarz, Witold Piekoszewski, Marian Szczerek, Jan Wulczyński, Massimo Lorusso, Jun Cao, Liang Qin, Aibing Yu, Guoping Li, Chul-Hee Lee, Peng Zhang, Chenglong Lian, Kwang-Hee Lee, Syed Mohammad Hasan Ahmer, Said Lal Jan, Nusratullah Khan, S Inayat Ali Shah, Ronaldo Cozza, Amaya Igartua, Raquel Bayon, Ana Aranzabe, Javier Laucirica

First published in London, United Kingdom, 2019 by IntechOpen
IntechOpen is the global imprint of INTECHOPEN LIMITED, registered in England and Wales, registration number: 11086078, 7th floor, 10 Lower Thames Street, London, EC3R 6AF, United Kingdom
Printed in Croatia

British Library Cataloguing-in-Publication Data
A catalogue record for this book is available from the British Library

Additional hard and PDF copies can be obtained from orders@intechopen.com

Friction, Lubrication and Wear
Edited by Mohammad Asaduzzaman Chowdhury
p. cm.
Print ISBN 978-1-78984-287-6
Online ISBN 978-1-78984-288-3
eBook (PDF) ISBN 978-1-83962-346-2

For EU product safety concerns:
IN TECH d.o.o., Prolaz Marije Krucifikse Kozulić 3, 51000 Rijeka, Croatia, info@intechopen.com or visit our website at intechopen.com.

Meet the editor

Mohammad Asaduzzaman Chowdhury is a professor of Mechanical Engineering at Dhaka University of Engineering and Technology (DUET), Gazipur, Bangladesh. His research interests include engineering tribology, surface engineering, coating technology, polymer and composite materials, metals and alloys, erosive wear of different materials, material fabrication, characterization of materials, self-energy-generating materials, and atomic transfer mechanism, among others. He serves as an editorial board member for many international journals of repute and is an editor and reviewer for many indexed journals and books. He has also published many research and review papers in refereed international journals and conference proceedings. He works as a consultant, advisor and expert member of many government and autonomous organizations. He has 20 years' experience as a teacher and researcher. He is involved with different cultural and social activities, such as writing articles, stories, lyrics, and poems.

Contents

Preface

The function of numerous electrical, electrochemical, mechanical, and biological systems or components depends on suitable friction and wear as well as tribological values. In this context, the study of friction, wear and lubrication is of tremendous pragmatic importance. In recent years, the study of tribological behavior of a component has received great attention, as the wastage of resources resulting from friction and wear has become clearer.

This book provides the motivation for research aimed at developing a fundamental understanding of the nature and consequences of the interaction between materials or components on friction, lubrication and wear perspectives. This book guides the rational design of material for technological application.

Chapter 1 "Abrasive Wear Performance of Fe2B Layers Applied on Steel Substrates" evaluates the resistance of dry abrasive wear in the Fe2B layer deposited on AISI D2 and 1040 steel substrates, using the powder-pack boriding process.

Chapter 2 "Experimental Results of the Tribology of Aluminum in the Presence of Polytron Additive" discusses experimental studies on a brand-new additive called Polytron to assess its role in the minimization of friction and wear.

Chapter 3 "Structural-Energy Interpretation of the Friction" studies the structural-energy model of elastic-plastic deformation as the main mechanism of transformation and dissipation of energy under friction. It proposes equations of friction energy balance, the energy interpretation of the coefficient of friction, and a structural-energy diagram of the friction surfaces.

Chapter 4 "Thin Films: Study of the Influence of the Micro-abrasive Wear Modes on the Volume of Wear and Coefficient of Friction" studies the influence of micro-abrasive wear modes on the behaviors of the volume of wear (V) and of the coefficient of friction (μ) of thin films submitted to micro-abrasive wear.

Chapter 5 "Novel Predictors for Friction and Wear in Drive Train Applications" derives a molecular model that shows how the release and the approach of additives toward a surface is essential and related to the reaction processes that occur during loading.

Chapter 6 "Tribological and Wear Behavior of Metal Alloys Produced by Laser Powder Bed Fusion (LPBF)" investigates the different behavior of principal metallic alloys by LPBF. Since LPBF is an additive manufacturing technique for the production of parts with complex geometry it is especially appropriate for structural applications in aircraft and automotive industries.

Chapter 7 "Tribological Characteristics of Smart Materials (Magneto-Rheological Fluids and Elastomers) and their Applications" shows the tribological characteristics of magneto-rheological fluids (MRFs) and magneto-rheological elastomers (MREs), as smart materials have been widely studied in various engineering fields to address vibration issues because the mechanical properties are controllable under the strength of a magnetic field.

Chapter 8 "A Review of Surface Treatments for Sliding Bearings Used at Different Temperature" reviews different treatment technologies to improve tribological properties and select the best treatment for different bearings used at different temperature. This is helpful considering that boundary lubrication and dry friction of plain bearings at different work are unable to be avoided under the start and stop conditions.

Chapter 9 "Tribology: The Tool to Design Materials for Energy-Efficient and Durable Products and Process" describes a summary of the main tribological achievements carried out at TEKNIKER technology centre during the last 37 years. It covers the description of commercial and newly developed tribological test benches and case studies for a wide variety of applications.

Chapter 10 "High-Temperature Self-lubricating Metal Nitride-Based Nanostructure Composite Films" reviews the formation mechanism of Magnéli phases and the influence of Magnéli phases on the high-temperature tribological properties of hard transition metal nitride films synthesized using magnetron sputtering. Film technology is one of the most efficient methods of modifying the surface properties of materials.

Chapter 11 "Testing of the Resistance to Scuffing of Spiral Bevel Gears - Test Rig, Method and Results of Verification Testing" studies the development of a new bevel gear test rig and scuffing test method at ITeE-PIB research institute. It investigates the effect of various gear oils and the deposition of a low-friction coating on the resistance to scuffing.

I am very much grateful to Dhaka University of Engineering & Technology (DUET), Gazipur, for creating the opportunities to do this work.

I wish to give my heartfelt and sincere thanks to Md. Bengir Ahmed Shuvho, Lecturer, National Institute of Textile Engineering and Research (NITER), Savar, Dhaka, for his continuous technical and valuable support. I appreciate and acknowledge his contributions in reviewing all the chapters. I immensely benefited from the help of Author Service Manager Ms. Ivana Barac and my publisher IntechOpen. I am also grateful to the authors for their worthy contributions. I learned a lot of ideas for editing this book from Professor Dr. Mohammed Alauddin, Honorable Vice Chancellor of DUET, Gazipur. Hopefully this book will be used for fundamental and applied research by students and researchers alike.

Special thanks go to my wife Anisa Akhtar, my son Ayan and my daughter Afsheen, who were forbearing during the year I spent long days and night in conducting research and keeping up with the literature as I prepared this book.

Mohammad Asaduzzaman Chowdhury
Dhaka University of Engineering and Technology,
Gazipur,
Bangladesh

Section 1

Friction and Wear

Chapter 1

Abrasive Wear Performance of Fe_2B Layers Applied on Steel Substrates

Armando Irvin Martínez Pérez, Edgar Ernesto Vera Cárdenas, Manuel Vite Torres, José Luis Bernal Ponce, Karina Alemán Ayala and Marisa Moreno Rios

Abstract

The resistance of dry abrasive wear in Fe_2B layer deposited on AISI D2 and 1040 steel substrates, using the powder-pack boriding process, was evaluated. The boriding process was carried out at temperatures of 1220 and 1320 K for a time of 8 h. A Rockwell hardness tester was used to assess the Daimler-Benz adhesion test. The abrasive wear tests were carried out in dry conditions according to the ASTM G65 test standard. The test parameters used were a sand flow of 400 g/min, a nominal rubber wheel constant rotation of 200 rpm, a load of 122 N, and a sliding distance of 716.28 m. The type of abrasive used was steel round grit with a grain size of 260 μm and a hardness of 1100 HV. The total time for each test was 30 min, removing the specimens every 5 min to determine the amount of mass loss using an analytical balance (sensitivity of 0.0001 g). The average value of volume loss and wear rates is reported. Optical microscopy and SEM were carried out in order to identify the wear mechanisms. The wear mechanisms presented in this study were two-body abrasive wear, pitting action, and plastic deformation.

Keywords: dry abrasive wear, Fe_2B layer, steel substrates, boriding, wear mechanisms

1. Introduction

Abrasive wear occurs when a hard particle slides on a surface, causing loss of material. This type of wear depends on factors such as hardness, roughness, and particle geometry [1–4].

Different coatings are used as anti-abrasive wear materials. Some of them are as follows: ceramics coatings, such as, Al_2O_3/TiO_2, $SiO_2/TiO_2/Cr_2O_3$, SiC, B_4C, ZrO_2, CaO, CrN/AlCrN, CrN/BCN, SiO_2, WC, and TiC [5, 6]; polymer coatings [7, 8]; and DLC coatings [9, 10].

On the other hand, some works have been developed using boron coatings as anti-wear material. Boronizing is a thermo-diffusion process in which boron atoms, due to their small diameter and high mobility at elevated temperatures, diffuse into a metal surface and form intermetallic compounds with atoms of base metal [11]. Abrasive wear tests were carried out using boronizing on SAE 1010, 1040, D2, and 304 steels [12]. It was seen that boronizing improved the wear strengths considerably. The best abrasive wear strengths were obtained in boronizing for 8 h at 900°C for SAE 1010 and SAE 1040 steels, 4 h at 900°C for D2 steel, and 6 h at 900°C for 304 steel. In another work [13], abrasive wear resistance of boride layers on Fe-15Cr alloy was studied. It was found that the dry abrasive wear resistance of borided alloy samples was around 45 times greater than that of non-borided ones. In another study [14], the micro-abrasive wear of boride layers on AISI D2 tool steel was investigated. Some results indicated that wear resistance of the borided samples was superior to the hardened, uncoated AISI D2 steel. According to literature [15], wear resistance of boronized steels in abrasive wear conditions depends on the phase composition and hardness of the layer and its stress state, but the hardness of abrasive particles also has a significant importance on the wear speed.

The objective of this work was to evaluate the resistance of dry abrasive wear in Fe_2B layer deposited on AISI D2 and 1040 steel substrates without a previous heat treatment (hardened and tempered) using the powder-pack boriding process. The substrate materials were selected in order to compare the wear abrasion behavior of a plain carbon (1040) versus a high-carbon, high-chromium steel (D2). AISI 1040 steel is frequently cold drawn to specified physical properties for use without heat treatment for some practical applications such as cylinder head studs. AISI D2 tool steel has desirable properties such as abrasion resistance, high hardness, and no deforming characteristics, and is used in lamination and stamping dies, shear blades, master tools, etc. The wear resistance of D2 tool steel is approximately eight times that of plain carbon steels, so also was of interest to know if this great difference in wear behavior could increase or decrease in borided conditions.

2. Experimental work

2.1 Test specimens

The specimens had a rectangular shape with dimensions of 50 × 25 mm and 10 mm in thickness. The chemical composition of the AISI D2 and 1040 steels is shown in **Table 1** [16, 17].

The boriding process used in the specimens of AISI D2 and 1040 steel substrates was the same as reported in a previous work [18]. The only difference is that for this work, two boriding temperatures were employed (1220 and 1320 K).

It is important to mention that, based on the Fe-B phase diagram and the high iron content of the AISI D2 and 1040 steels [19], in addition to the diffusion of boron at high temperatures (1220 and 1320 K) and the treatment time of 8 h, the formation of a Fe_2B monolayer is ensured under the conditions of the boriding process proposed in this work. Because the Fe-B phase is formed on the surface of the sample, which generates the Fe-B/Fe interface between the Fe-B phase and the steel, this allows the gradual formation of the Fe_2B phase that grows when the thickness of the boride is increased and at the same time the Fe-B phase decreases. At the

Steel	Composition								
	C	Mn	Si	Cr	Mo	V	S	P	Fe
AISI D2	1.55	0.35	0.35	11.8	0.85	0.85	0.03	0.03	84.22
AISI 1040	0.38	0.6	0.1	0	0	0	0.02	0.02	98.87

Table 1.
Chemical composition of specimens (wt.%).

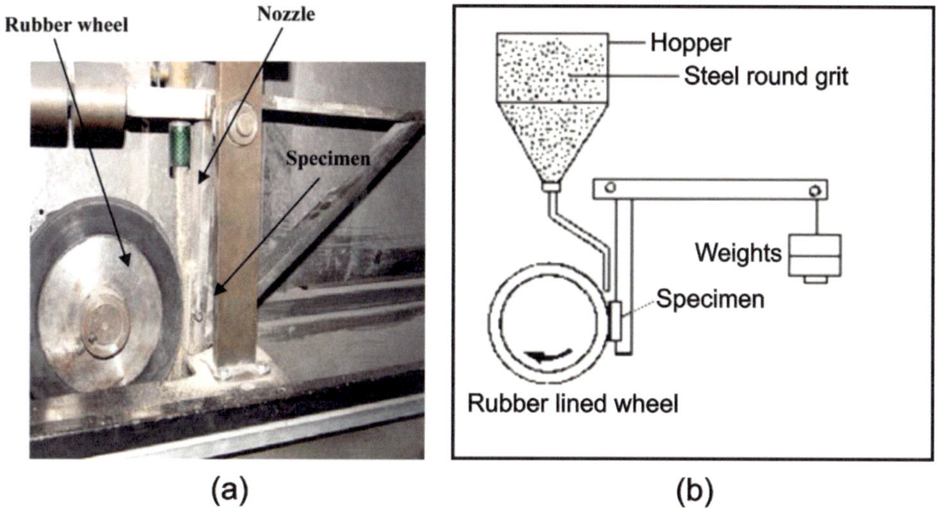

Figure 1.
(a) Experimental setup and (b) schematic diagram of the apparatus.

end of the 8 h, corresponding to the treatment time, the Fe-B phase was consumed completely, and so the only phase present (Fe_2B) stops growing [20].

2.2 Test procedure

The tests were performed according to the ASTM G65 test standard [21]. **Figure 1** shows the experimental rig and a simplified schematic diagram of the dry/sand rubber wheel apparatus used in this research work [22].

The test parameters used were a sand flow of 400 g/min, a nominal rubber wheel constant rotation of 200 rpm, a load of 122 N, and a total sliding distance of 716.28 m, using a 228.6 mm diameter wheel rotating. The wheel consists of a steel disk with an outer layer of neoprene rubber tire molded to its periphery with hardness A60. As the rubber wheel reduces in diameter, the number of wheel revolutions was adjusted to equal the sliding distance of the new wheel. The type of abrasive used was steel round grit with a grain size of 260 μm and a hardness of 1100 HV. The total time for each test was 30 min, removing the specimens every 5 min to determine the amount of mass loss using an analytical balance (sensitivity of 0.0001 g). Before the overall tests were performed, the specimens were cleaned by washing in ethanol in an ultrasonic bath (Fisherbrand 11020).

The average value of volume loss (V), wear rates (Q), and wear coefficients (k) are reported. Optical microscopy and SEM were carried out on the damaged surfaces in order to identify the wear mechanisms. Additionally, the profiles of the wear scars are presented using a Mitutoyo Surftest Profilometer.

3. Results and discussion

3.1 Fe_2B layer hardness

A load of 100 g was used to evaluate the hardness of Fe_2B layers with a Vickers indenter. The variation of the hardness, depending on the depth of layers, is shown in **Figure 2**. Also, the roughness of specimens was obtained with a Mitutoyo Surftest Profilometer, see **Table 2**.

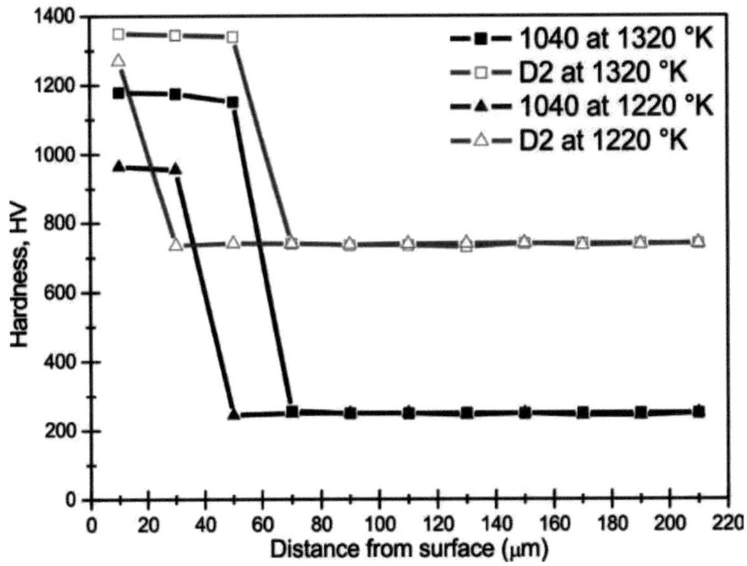

Figure 2.
Variation of hardness.

Boriding temperature (K)	Borided steel	Vickers hardness (HV)	Roughness (Ra) (μm)
1220	AISI D2	1270.7	0.86
	AISI 1040	964.4	1.14
1320	AISI D2	1354.5	0.22
	AISI 1040	1179.5	0.35

Table 2.
Properties of the specimens.

3.2 SEM, X-ray diffraction, and EDS

Figure 3 shows the cross-sectional view of SEM micrographs. A zigzag teeth shape is observed in both steels. This columnar shape comes from the direction in which diffusion is preferred, and the boride is of stronger (002) texture [23]. The presence of this typical morphology for good adhesion between coating and substrate is necessary.

The boriding is a diffusive process highly anisotropic [24]. In **Figure 3**, it was observed that the boride on the surface of AISI D2 steel presents a columnar morphology; in the case of boride formed on the surface of AISI 1040 mold steel is observed a dense structure due to alloying elements it has. Depending on the conditions of processing time, temperature, and chemical composition of substrates, the depth obtained of the boride layer was an interval of 10–60 μm (**Figure 2**). It was observed that the depth of borides formed on AISI D2 is more homogeneous than that of AISI 1040.

The results of X-ray diffraction studies are presented in **Figure 4**. The XRD analysis shows well-defined peaks at 42.67 and 45.11° confirming Fe_2B phase. Also, the presence of chromium boride (CrB) phase in the borided AISI D2 steel was determined. This is due to the significant presence of chromium in AISI D2 steel as an alloying element [25, 26]; apparently, during powder-pack boriding, it reacted with boron atoms and formed a little intermediate phase of CrB.

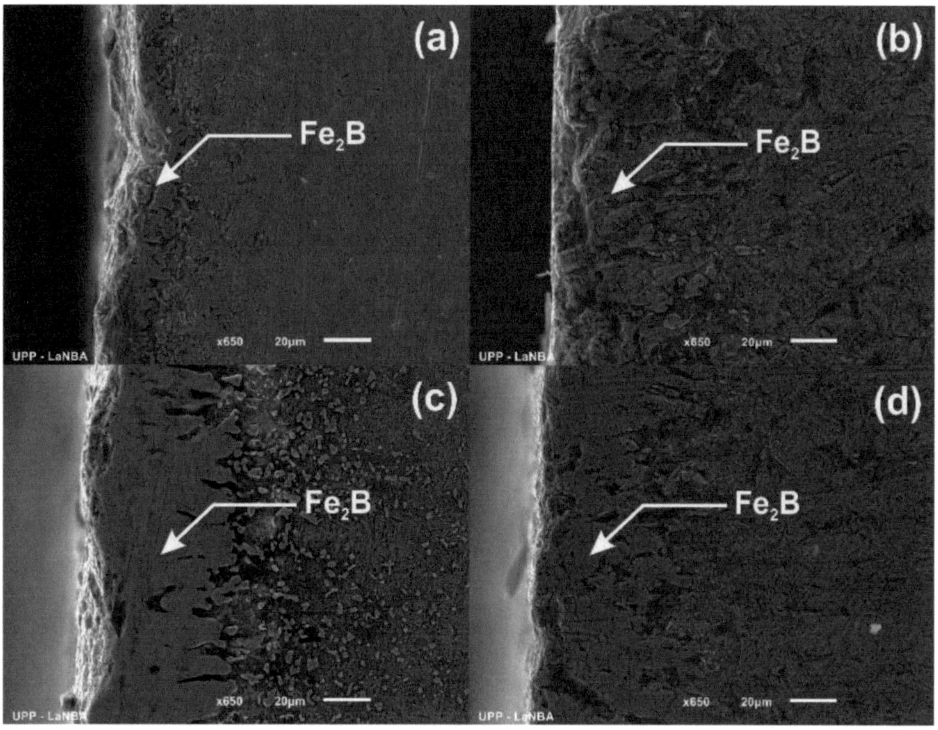

Figure 3.
SEM cross-sectional micrograph, and XRD of borided samples: (a) AISI D2 at 1220 K, (b) AISI 1040 at 1220 K, (c) AISI D2 at 1320 K and (d) AISI 1040 at 1320 K.

The EDS analysis obtained by SEM, for the borided steels, is shown in **Figure 5a–d**. The presence of borides formed on the surfaces of the steels was confirmed considering the presence of boron and iron.

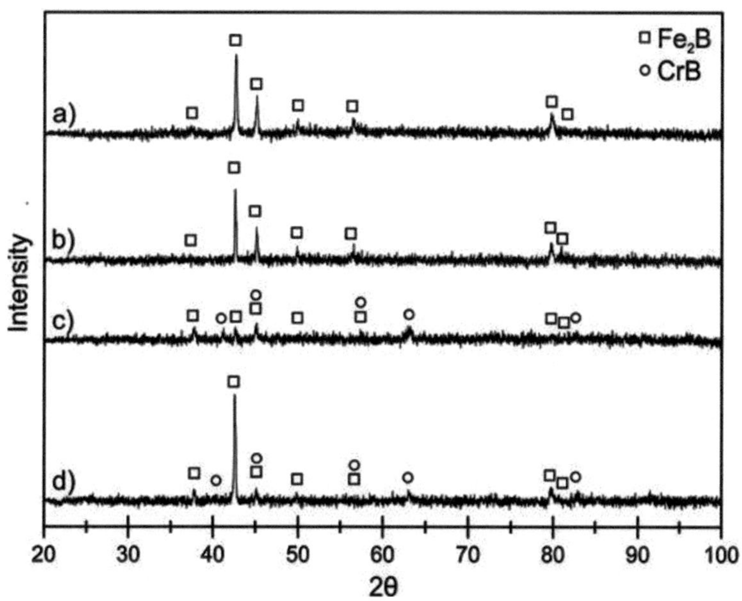

Figure 4.
Diffraction patterns of borided specimens: (a) AISI 1040 at 1220 K, (b) AISI 1040 at 1320 K, (c) AISI D2 at 1220 K, and (d) AISI D2 at 1320 K.

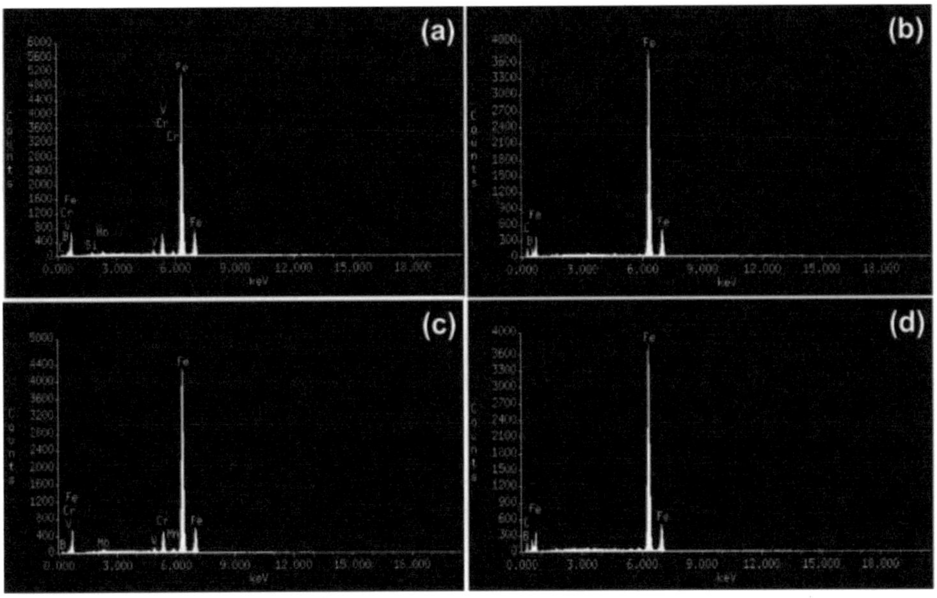

Figure 5.
EDS spectrum of borided samples. (a) AISI D2 at 1220 K, (b) AISI 1040 at 1220 K, (c) AISI D2 at 1320 K, and (d) AISI 1040 at 1320 K.

3.3 Fe_2B layer adhesion test

A Cientec Rockwell hardness tester model 200HR-150 was used to assess the Daimler-Benz adhesion tests [27]. **Figure 6** shows the indentations on the surfaces. For the AISI D2 steel, some small cracks and no visible delamination are observed (**Figure 6a** and **c**), and the adhesion strength quality is related to HF1 map [28]. In the case of AISI 1040 steel (**Figure 6b** and **d**), microcracks and small delamination are observed, and the adhesion category belongs to the HF4 level.

3.4 Wear profile

The abrasion tests carried out caused wear damage on surfaces. The wear profiles were measured using a Mitutoyo Surftest profilometer and are shown in **Figure 7**. The results are compatible with the volume loss (**Figure 8**), where 1040 steel borided at 1220 K has the greatest wear volume and the D2 steels at 1320 K had the minor wear.

3.5 Volume loss

The volume loss was obtained for all the borided and unborided steel substrates. These data were calculated using Eq. (1). The mass loss was obtained weighing the specimens before and after the test. The graph of the **Figure 8** shows that the AISI D2 steel borided at 1320 K exhibited a higher wear resistance compared to the other specimens. The results also show the great difference in volume loss between borided and unborided steels. The reason that the D2 steel had a greater wear resistance is due most likely to the mechanical properties conferred by a high content of C and Cr, which is higher than in the 1040 steel (see **Table 1**). According to **Figure 8**, the abrasion wear resistance of borided D2 tool steel is approximately 16 times greater than borided 1040 plain carbon steel. This could justify the use of this tool steel for abrasion wear applications when it is borided to the conditions used in this work,

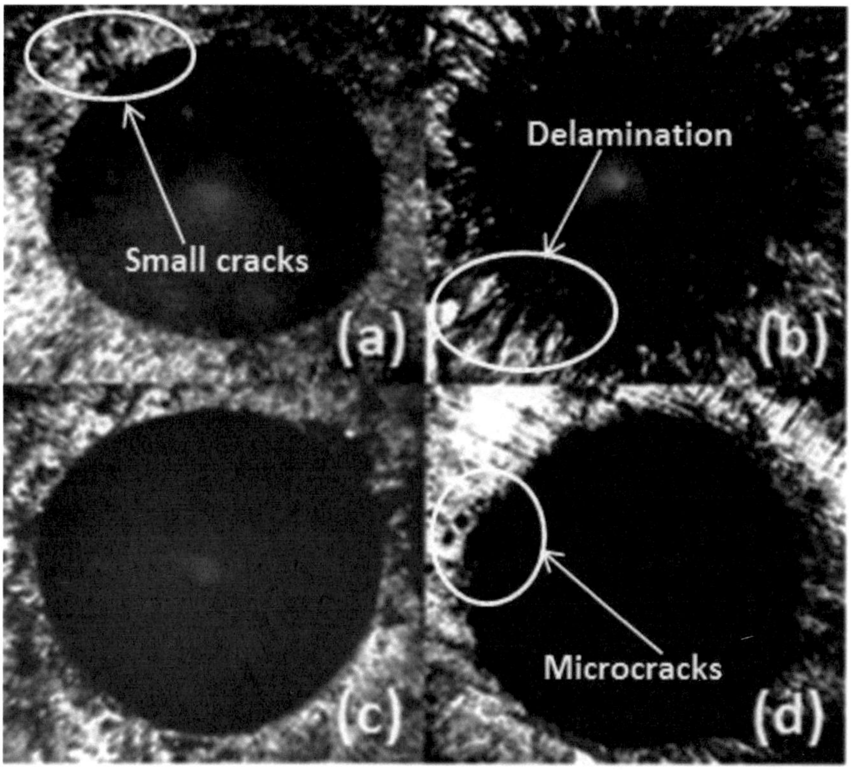

Figure 6.
Indentations on surfaces. (a) AISI D2 borided at 1220 K, (b) AISI 1040 borided at 1220 K, (c) AISI D2 borided at 1320 K, and (d) AISI 1040 borided at 1320 K.

without a previous heat treatment, as usual. These results also indicate that its core strength was not affected due to high temperatures of powder-pack boriding treatment.

$$\text{Volume loss}\left(\text{mm}^3\right) = \left(\text{Mass loss}\,(\text{g})/\text{density}\left(\text{g/cm}^3\right)\right) \times 1000 \qquad (1)$$

3.6 Wear rates (Q) and wear coefficients (k)

The wear rates were obtained from Eq. (2).

$$Q = V/d. \qquad (2)$$

where Q = wear rate (mm^3/m), V = volume loss (mm^3), and d = sliding distance (m).

Figure 9 shows the wear rates obtained every 716 m of the borided and unborided D2 and 1040 steels. AISI D2 borided steels at 1320 and 1220 K had the best performance against the dry abrasive wear conditions. It was due to its good mechanical properties and chemical composition. Additionally, the adhesion tests on this steel showed an excellent performance. On the other hand, the 1040 steel borided at 1320 K had a good behavior, almost similar to the D2 steel. For the AISI 1040 steels borided at 1220 K, an abnormal value was observed at a sliding distance of 716 m, where the wear rate had a great increase. This performance was mainly due to the running in period of the test.

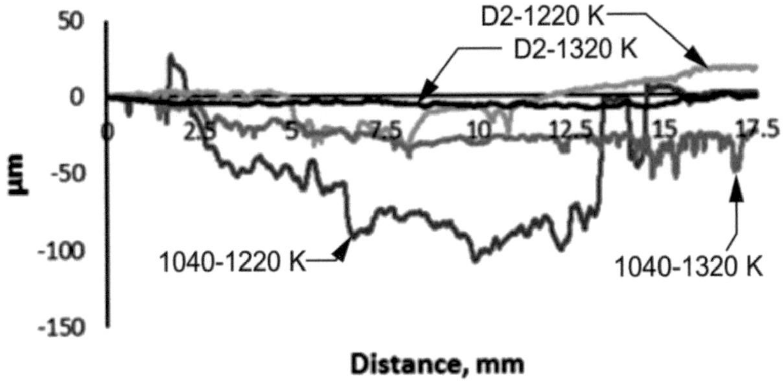

Figure 7.
Roughness profiles of wear scars.

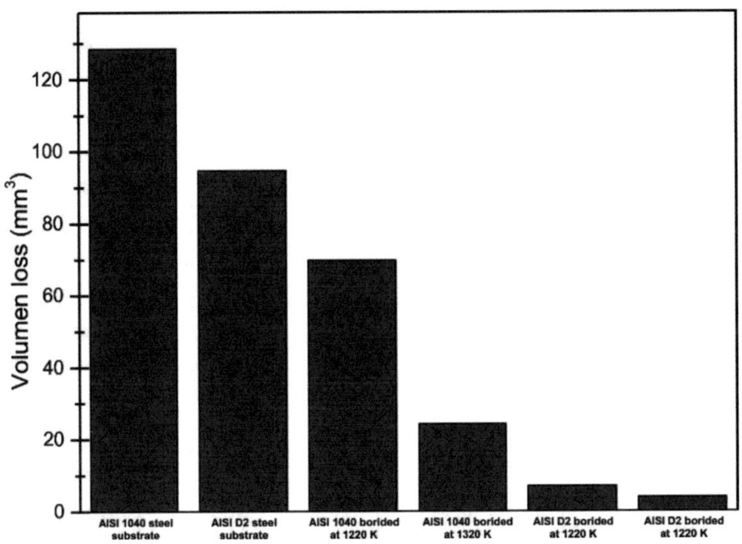

Figure 8.
Volume loss of borided and unborided steels.

3.7 Wear mechanisms

Figure 10a–f shows the wear on the surface of the unborided and borided specimens. In both steels, as expected, the wear damage was more severe on the unborided specimens (**Figure 10a** and **b**). In the case of the borided steels, AISI 1040 at 1220 K (**Figure 10c**) showed the greater damage and the AISI D2 borided at 1320 K (**Figure 10d**) showed the lower damage. This was in accordance to the results of the wear profiles showed in **Figure** 7 and wear rates showed in **Figure 9**. In the case of borided steels, the main wear mechanism observed in the wear scars was the two-body abrasive wear due to the presence of parallel lines to the sliding direction. These parallel lines were produced by wear debris acting as indenters causing, in some cases, depth grooves.

Figure 11a–d shows the SEM micrographs of the wear damage of the borided steels. In the case of AISI 1040 steel borided at 1220 K (**Figure 11a**), wear debris was observed, which was derived from the three-body abrasion situation, where hard particles were trapped between the two sliding surfaces. Also, severe pitting action was observed, caused by the particles of material and abrasive particles. Finally,

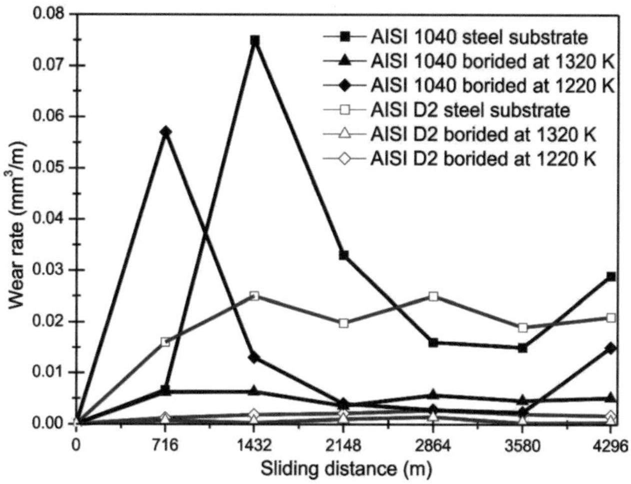

Figure 9.
Wear rate of borided and unborided steels.

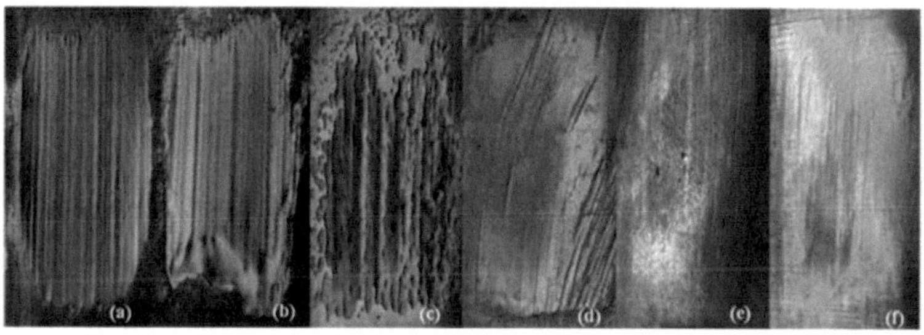

Figure 10.
(a) 1040 steel substrate, (b) D2 steel substrate (c) 1040 borided—1220 K, (d) D2 borided—1220 K, (e) 1040 borided—1320 K, and (f) D2 borided—1320 K.

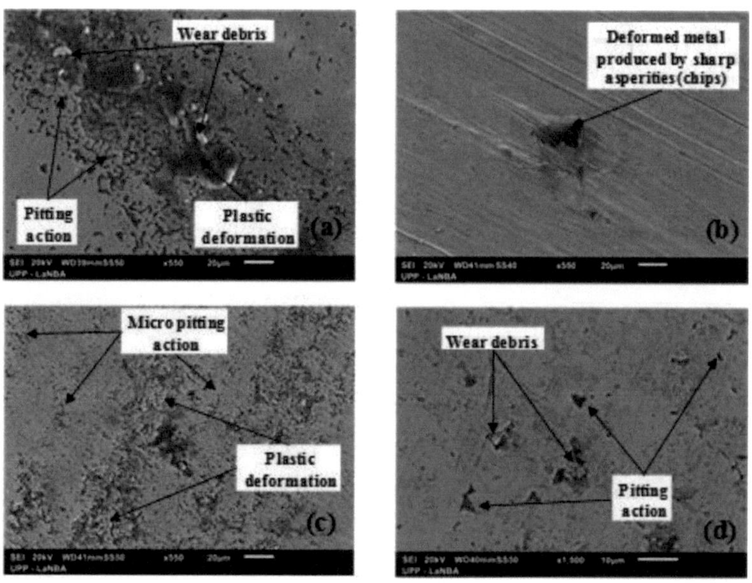

Figure 11.
SEM micrographs of wear scars. (a) AISI 1040 at 1220 K, (b) AISI D2 at 1220 K, (c) AISI 1040 at 1320 K, and (d) AISI D2 at 1320 K.

plastic deformation occurred due mainly to the sliding action. **Figure 11b** (AISI D2 steel at 1220 K) shows a typical effect of abrasive wear, where sharp asperities plowed the surface, causing permanent deformation. This plowing action produced a groove in the softer material. Also, parallel lines were observed caused by wear debris probably inlaid in the rubber wheel. In the case of AISI 1040 steel borided at 1320 K (**Figure 11c**), severe micropitting action and plastic deformation were observed. Finally, in the case of the AISI D2 steel at 1320 K, wear debris and pitting action were observed, but to a lesser degree. This matches with the results of wear rates obtained and indicated in **Figure 9**.

4. Conclusions

1. The formation of a Fe_2B monolayer is ensured under the conditions of the process of boron powder-pack proposed in this work. The Fe-B phase is formed on the surface of the sample, which generated the Fe-B/Fe interface between the Fe-B phase and the steel, which allowed the gradual formation of the Fe_2B phase that grows when the thickness of the boride is increased and at the same time the Fe-B phase decreases.

2. The Rockwell-C adhesion tests showed that for the AISI D2 steel, the adhesion strength quality, of Fe_2B layers, is related to HF1 map, showing some small microcraks and the AISI 1040 steel fits to HF4 category, where microcracks and some delamination was observed.

3. AISI D2 steel specimens borided at 1320 K showed a higher wear resistance, in accordance to the wear rates and wear coefficients results. It was due to its good mechanical properties and chemical composition. Additionally, the adhesion tests on this steel showed an excellent performance.

4. The wear mechanisms presented in this study were as follows: two-body abrasive wear, which was due to the presence of parallel lines to the sliding direction; pitting action, which was caused by the particles of the material and abrasive particles; and plastic deformation, which occurred due mainly to the sliding action.

Author details

Armando Irvin Martínez Pérez[1], Edgar Ernesto Vera Cárdenas[2]*, Manuel Vite Torres[3], José Luis Bernal Ponce[4], Karina Alemán Ayala[5] and Marisa Moreno Rios[2]

1 Polytechnic University of Pachuca, Zempoala, Hidalgo, Mexico

2 National Technology of Mexico/Technological Institute of Pachuca, Pachuca, Hidalgo, Mexico

3 National Polytechnic Institute, ESIME Zacatenco, Mexico City, Mexico

4 National Technology of Mexico/Technological Institute of Orizaba, Orizaba, Veracruz, Mexico

5 Autonomous University of the State of Hidalgo, Mineral de la Reforma, Hidalgo, Mexico

*Address all correspondence to: eevc2000@hotmail.com

References

[1] Rabinowicz E. Friction and Wear of Materials. New York, USA: Wiley; 1965

[2] Bhushan B. Introduction to Tribology. New York, USA: Wiley; 2002

[3] Khruschov MM. Principles of abrasive wear. Wear. 1974;**28**:69-88

[4] ASTM G40-17. Standard Terminology Relating to Wear and Erosion. 2017

[5] Mo JL, Zhu MH. Sliding tribological behaviors of PVD CrN and AlCrN coatings against Si_3N_4 ceramic and pure titanium. Wear. 2009;**267**: 874-881

[6] Czerwinski F. Thermochemical Treatment of Metals. Croatia: Intech; 2012

[7] Zhu F, Wang J, Li S, Zhang J. Preparation and characterization of anodic films on AZ31B Mg alloy formed in the silicate electrolytes with ethylene glycol oligomers as additive. Applied Surface Science. 2012;**258**:8985-8990

[8] Brzeziński S, Kowalczyk D, Borak B, Jasiorski M. Applying the sol–gel method to the deposition of nanocoats on textiles to improve their abrasion resistance. Journal of Applied Polymer Science. 2012;**125**:3058-3067

[9] Corbella C, Rubio-Roy M, Bertran E, Polo MC, Pascual E, Andújar JL. Low friction and protective diamond-like carbon coatings deposited by asymmetric bipolar pulsed plasma. Diamond and Related Materials. 2009;**18**:1035-1038

[10] He F, Wong PL, Zhou X. Wear properties of DLC-coated steel rollers running with highly contaminated lubrication. Tribology International. 2010;**43**:990-996

[11] Milinović A, Krumes D, Marković R. An investigation of boride layers growth kinetics on carbon steels. Tehnički vjesnik. 2012;**19**:27-31

[12] Atik E, Yunker U, Meric C. The effects of conventional heat treatment and boronizing on abrasive wear and corrosion of SAE 1010, SAE 1040, D2 and 304 steels. Tribology International. 2003;**36**:155-161

[13] Dybkov VI, Goncharuk LV, Khoruzha VG, Samelyuk AV, Sidorko VR. Growth kinetics and abrasive wear resistance of boride layers on Fe–15Cr alloy. Materials Science and Technology. 2011;**27**:1502-1512

[14] Oliveira CK, Lombardi AN, Totten GE, Casteletti LC. Micro-abrasive wear of boride layers on AISI D2 tool steel produced by the thermoreactive process. International Journal of Microstructure and Materials Properties. 2008;**3**:241-253

[15] Krukovich MG, Prusakov BA, Sizov IG. Plasticity of Boronized Layers. Russia: Springer; 2016. pp. 250-253

[16] ASTM A681-08. Standard Specification for Tool Steels Alloy; 2015 (Reapproved)

[17] ASTM A510-03. Standard Specification for General Requirements for Wire Rods and Coarse Round Wire, Carbon Steel. 2003

[18] Vera Cardenas E, Lewis R, Martinez Perez A, Bernal Ponce J, Perez Pinal F, Ortiz Dominguez M, et al. Characterization and wear performance of boride phases over tool steel substrates. Advances in Mechanical Engineering. 2016;**8**(2):1-10

[19] Repovský P, Homolová V, Čiripová L, Kroupa A, Zemanová A. Experimental study and thermodynamic modelling of the

B-Fe-Mn ternary system. CALPHAD: Computer Coupling of Phase Diagrams and Thermochemistry. 2016;**55**:252-259

[20] Yu LG, Chen XJ, Khor KA, Sundararajan G. Fe-B/Fe_2B phase transformation during SPS pack-boriding: Boride layer growth kinetics. Acta Materialia. 2005;**53**:2361-2368

[21] ASTM G65-04. Standard Test Method for Measuring Abrasion Using the Dry Sand/Rubber Wheel Apparatus. 2010

[22] Vite Torres M, Moreno Rios M, Gallardo Hernandez E, Laguna Camacho J. A study of the abrasive resistance of sputtered CrN coatings deposited on AISI 316 and AISI H13 steel substrates using steel particles. Wear. 2011;**271**:1273-1279

[23] Martini C, Palombarini G, Carbucicchio M. Mechanism of thermochemical growth of iron borides on iron. Journal of Materials Science. 2004;**39**:933-937

[24] Uslu I, Comert H, Ipek M, Celebi FG, Ozdemir O, Bindal C. A comparison of borides formed on AISI 1040 and AISI P20 steels. Materials and Design. 2007;**28**:1819-1826

[25] Aronsson B, Aselius J. The effect of boron on the formation of α-FeCr at 700°C. Acta Chemical Scandinavica. 1958;**12**:1476-1480

[26] Nedfors N, Primetzhofer D, Wang L, Lu J, Hultman L, Jansson U. Characterization of magnetron sputtered Cr–B and Cr–B–C thin films for electrical contact applications. Surface and Coatings Technology. 2015;**266**:167-176

[27] Taktak S, Tasgetiren S. Identification of delamination failure of boride layer on common Cr-based steels. Journal of Materials Engineering and Performance. 2006;**15**:570-573

[28] Verein Deutscher Ingenieure Normen VDI 3198. Düsseldorf: VDI-Verlag (Coating (CVD, PVD) of cold forging tools); 1991. pp. 1-8

Chapter 2

Experimental Results of the Tribology of Aluminum in the Presence of Polytron Additive

Syed Mohammad Hassan Ahmer, Nusratullah Khan, S. Inayat Ali Shah and Lal Said Jan

Abstract

Friction is an ever-present obstacle that causes energy loss in mechanical parts. To alleviate this nuisance, we carried out experimental studies on a brand new additive called Polytron to assess its role in the minimization of friction and wear. The wear, the volume wear rate, the wear coefficient, and the coefficient of friction of the aluminum surface were measured at room temperature with pin-on-disk tribometer without and with 10% Polytron in Helix oil. In the base oil Helix, their values were found to be 70 μm, 1.28×10^{-3} mm^3/min, 1.27×10^{-10} m^2/N, and 0.012, respectively, which with the incorporation of Polytron additive in the Helix oil correspondingly reduced to 20μm, 6.08×10^{-5} mm^3/min, $4.22 \times 10^{-11} \frac{m^2}{N}$, and 0.004. The experimental verdict points to an ionic character of the additive in that it impregnates the crystal structure of the metal, thereby prompting a hard surface layer which subsequently curtails wear and friction.

Keywords: friction, wear rate, polytron additive, aluminum metal, lubrication, helix oil

1. Introduction

Whenever and wherever two surfaces and/or two parts move against each other in the form of translation, rotation, or oscillation, an opposition is encountered. This opposing or resistive force to motion is described as friction. In fact, friction is an ever-present irritant and is the real source of energy and power losses in every industry and every activity whatsoever. This can be realized in our everyday life and the different industries like automotive, aerospace, agriculture, marine, electronics, and telecommunication, and even the so-called cosmetics industry, and the movements of the human joints are not exempt from this scourge one way or another. The word friction derives from the Latin verb fricare, which means to rub. It is of interest to know that the word tribology, introduced in 1966 by the Jost Report, derives from the Greek word τριβοσ (tribos), which also means rubbing. As indicated by this report, tribology was defined as the science and technology of interacting surfaces in relative motion. Nevertheless, a better definition of tribology might be the science and technology of lubrication, friction, and wear of moving or stationary parts [1, 2]. Even if the term tribology is difficult for the general public to comprehend, the dawn of computer disk drives, micro-devices, and nanotechnology has driven friction science and tribology to the front position. Now, the designers have to deal with the challenge of controlling friction of interacting surfaces in relative motion at sizes far too small for the naked eye to see. This is the nano-mechanical device and nano-tribological regime where the ultimate source of friction is perceived to be van der Waals force and Coulomb force [3–7]. In addition to friction, an associated observable fact with the protracted mechanical motion or rubbing of the mating surfaces is the wreckage of the surfaces and generation of heat and pressure in the surrounding area which will definitely curtail the useful life of the mechanical parts. This scoring of the coupling surfaces is termed as wear. The critical issue is to minimize the amount of wear and friction being produced in any mechanical operation so as to avoid any possible mechanical malfunction. It is hard to stop wear of the surfaces and generation of heat and pressure; but there are different ways to minimize the effects, and one of them is lubrication [7]. A lubricant is any substance that is interposed between two surfaces in relative motion for the purpose of reducing the friction and wear between them. By and large, lubricants can be solids, liquids, or gases; but in any case, they reduce the negative influence in the moving parts. Other than friction reduction, lubricants carry away heat and wear particles as well and can serve as the means to distribute corrosion inhibitors and biocides. Lubricating films should support the pressure between opposing surfaces, separate them, and reduce the sliding or rolling resistance in the interface. To reduce friction, the liquid lubricants are formulated in such a way that chemical species within it react with the surface of the bodies to form lubricative films. This chemical species is named as additive. The function of the additive is to provide a smooth surface plus reduce the amount of wear; that is, they are expected to have antifriction and antiwear properties. For example, calcium sulfonate causes the formation of protective layers on highly loaded surfaces. Phosphorus can react with frictional hot spots on ferrous surfaces and thus can reduce wear and friction. Friction modifiers and antiwear additives to oils are the focus of extensive research in oil companies. The amount of the above-mentioned components and their nano-sized counterparts can vary, depending upon the application, in the range of 1–20 wt% [7–15]. By the same token, it has been noticed that the variation of friction and wear rate depends on various interfacial conditions. There are a number of studies in the literature which report that wear and friction primarily change with load, speed, and/or temperature [16–22], surface roughness [23, 24], type of material or mating component, and other environmental dynamics [25–30]. Yet, a group of researchers argue that friction and wear rate vary with geometry, relative surface motion, surface roughness of the rubbing surfaces, type of the material, system rigidity, stick-slip, lubrication, and vibration and/or type of additive, which means that wear and friction are functions of the specific tribosystem [31–52]. Even then, in many applications, the wear reduction mechanism and quantitative analysis of the additives are not well known and a thorough exploration is still inevitable. A literature survey reveals that there is a peculiar and unexplored additive with the brand name of Polytron which has not been thoroughly investigated by the tribological community. Accordingly, this chapter has been devoted to an academic research on the Polytron additive. Polytron is an oily fluid mixture of petroleum-based chemicals mixed with oxidation inhibitors and detergent chemicals and behaves exactly like a stable grease at ambient pressure and temperature in stark contrast to the conventional lubricants. Polytron additive is petroleum based and thus contains no solid particles; hence, it is compatible with all the lubricants available in the market whether mineral, synthetic, vegetable, or animal. Polytron comprises 80% para and 20% meta polytron. In this chapter, we will focus on the metal treatment concentrate (MTC) trademark of polytron having an inherent ionic/polar nature due to which it is attracted to metallic surfaces and develops a durable polished-like microscopic layer through metallurgical process that can resist wear, extreme pressure, and excessive temperature.

2. Experimental details

2.1 Tribometer machine

Wear tests were conducted on pin-on-disk tribotester (Ducom TR-20LE) wear testing machine. **Figure 1** gives a schematic sketch of the pin and disk while **Figure 2** displays the actual tribometer device.

In **Figure 1**, F_N stands for the normal force that is the load on the aluminum pin whereas F_R represents the resistive force called friction that arises from the sliding contact of the aluminum pin on the steel disk. In **Figure 2**, the pin is firmly attached to the pin support and then linked to the rotating plain disk with the desired load which is usually applied through a pulley system. Lubricant is pumped continuously from the machine. To simplify the contact geometry, a hemispherical pin is used which directly touches the disk surface at the beginning of the experiment. A hygrometer measures the relative humidity of the air in the chamber whereas the rpm of the rotating shaft that supports the disk is measured with the help of tachometer. The variation of friction coefficient with friction time is recorded automatically. Necessary information regarding stainless steel disk and aluminum pin is presented in **Tables 1–3**. The aluminum pin is in fact an alloy of aluminum and silicon. In addition, the data sheets for the Helix oil and Polytron additive are given in **Tables 4** and **5**. The data and basic information with reference

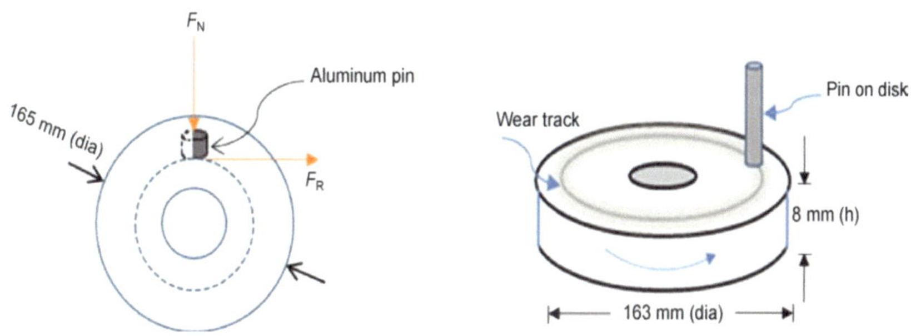

Figure 1.
Sketch of the pin-on-disk. The dimensions of the pin were 32 mm (l) × 10 mm(dia) and dimensions of the disk were 8 mm (l) × 163 mm (dia).

Figure 2.
Pin-on-disk tribotester machine.

Component	Specification/weight percent
Disk dimensions	165 nm (diameter) × 8 mm (height)
Counter bore	M5 holes from bottom × 4 nos.
Counter bore	M5 holes from top × 4 nos.
Holes	M4 tapped holes × 2 nos.
Chemical composition (weight percent)	
Carbon (C)	≤ 0.08%
Silicon (Si)	≤ 1.00%
Manganese (Mn)	≤ 2%
Phosphorous (P)	≤ 0.045%
Sulfur (S)	≤ 0.30%
Nickel (Ni)	≤ 8 –10.5%
Chromium (Cr)	≤ 18.00–20.00%

Table 1.
Specification and composition of stainless steel disk (SUS304) [53–55].

Composition	Min (weight percent)	Max (weight percent)
Silicon	0.4%	0.8%
Iron	—	0.7%
Copper	0.15%	0.15%
Manganese	—	0.15%
Magnesium	0.8%	1.2%
Chromium	0.04%	0.35%
Zinc	—	0.25%
Titanium	—	0.15%
Aluminum	95.85%	98.56%

Table 2.
Chemical composition of aluminum pin (A390) [53–55].

Property	Alloy	
	Aluminum pin A390	Steel disk SUS304
Density	2.72 g/cm^3	8000 kg/m^3
Hardness	112.65 VHN	88 HB
Tensile strength	250.00	520 MPa
Yield strength	—	240 MPa
Young's modulus	—	190 GPa
Poisson ratio	—	0.27–0.30

Table 3.
Mechanical properties of the aluminum pin and steel disc [53–55].

Property	Method	Shell Helix Ultra
SAE viscosity grade		5W–40
Kinematic viscosity		
@40 °C cSt	IP 71	81.1
@100 °C cSt	IP 71	14.5
Viscosity index	IP 226	187
Density @15 °C (kg/L)	IP 365	0.856
Flash point PMCC (°C)	IP 34	206
Pour point (°C)	IP 15	–39
HTHS viscosity @ 150°C (mPa s)		3.68

Table 4.
Typical physical properties of Shell Helix Ultra oil (5W–40) [53–55].

Physical/chemical property	Remarks
State	Liquid
Color	Yellowish clear
Smell	Odorless
Specific gravity	60/60 ≈ 1.00
Boiling point range	>300°C
Flash point	>200°C
Viscosity @100 °F	SUS 391
Viscosity @210 °F	SUS 61
Water solubility (T = 20 °C)	Low
Evaporation point	Higher than ether (>34.6°C)

Table 5.
Data sheet of Polytron [53–55].

to aluminum metal, steel disk, Helix oil, and polytron are taken from the research work of Ahmer et al. [53], John [54], and Ahmer et al. [55]. One can guess from **Table 5** that Polytron is marketed in a liquid state and is yellowish in color and, unlike other solid additives, it is odorless. Its flash point is beyond 200 °C whereas its boiling point is further than 300 ° C and it is scarcely soluble in water.

2.2 Materials and chemicals

The experimental work was performed in the Tribology Laboratory of Universiti Kebangsaan Malaysia, UKM,at ambient temperature (300 K) and pressure (760 mmHG) and approximately 70% relative humidity. Helix oil was chosen as representative base oil for the experiment and its brand 5W–40 was supplied by Shell Oils. The additive was Polytron MTC which was supplied by the Malaysian Association of Productivity. We used soft aluminum-silicon alloy A390 and stainless steel SUS304 as pin and disk material, respectively. Separate test runs were taken for the base oil stock and the 10% polytron additive plus the base oil stock. The runs were executed for 240 min in each case and the wear rates of the pin were then calculated from the measured weight loss. The mass and volume of the pin

Test variable	Assessed value
Before the wear run	
Material of the wear disk	Stainless steel S304
Diameter of the wear disk	80 mm
Mass of the pin	6.4480 g
Length of the pin	32.00 mm
During the wear run	
Speed of the wear disk	500 rpm
Time allocated	240 min≈ 14,400 s
Sliding speed	2.09 m/s
Sliding distance	30.163 km ≈ 30,163.2 m
After the wear run	
Mass of the pin	6.4470 g
Length of the Pin	31.981 mm

Table 6.
Recorded data of the wear test for helix base oil (5W–40).

Test variable	Assessed value
Before the wear run	
Quantity of Helix plus Polytron	2000 mL
Load	196.2 N
Material of the pin	Al■Si alloy A390
Pin diameter	10.00 mm
Length of the pin	32.00 mm
Material of the wear disk	Stainless steel SUS 304
Diameter of the wear disk	80 mm
During the wear run	
Speed of the wear disk	500 rpm
Time allocated	240 min≈ 14,400 s
Sliding speed	2.09 m/s
Sliding distance	30.163 km ≈ 30163.2 m
After the wear run	
Mass of the pin	6.4472 g
Length of the pin	31.996 mm

Table 7.
Recorded data of the wear test for the Helix oil plus 10% polytron.

were measured both before and after running the experiment and the data set are presented in **Tables 6** and 7.

2.3 Procedure and calculations

In the experiment, an aluminum pin having a diameter of 10 mm was slid against the steel disk. The applied load was 20.0 kg. In the first instance, 100%

Helix oil was used and its volume in the graduated cylinder was 2000 mL. In the second instance, 90% Helix oil having a volume of 1800 mL was mixed with 10% polytron additive which amounted to 200 mL volume of polytron. Before running the test, the disk was completely covered with the lubricant by keeping a steady flow rate of the lubricant at nearly 0.5 mL/min. The wear volume was calculated from the diameter of the wear scar generated by the pin. Typical wear versus time curves were obtained with the help of MatLab software and were polynomially fitted in order to decide the data trend.

Wear process is in general quantified by the wear rate. Wear rate is defined as the volume or mass of material removed per unit time or per unit sliding distance. In order to determine the extraordinary contribution of the polytron additive in the helix lubricant, we calculated three key tribological parameters, namely, mass wear rate, volume wear rate, and wear coefficient. The defining equations for these parameters are specified by Eqs. (1–3) as written down below [56].

$$\text{Mass wear rate} = m/t \tag{1}$$

$$\text{Volume wear rate} = V/t \tag{2}$$

$$\text{Wear coefficient (k)} = (V \times H)/(N \times S) \tag{3}$$

Eq. (3) is the famous Archard equation of tribology.

The coefficient of friction μ is obtainable from the experimentally obtained data. The popular defining expression for the coefficient of friction is like that described by Eq. (4).

$$\mu = F_R/N \tag{4}$$

In the above equations, the variable m stands for the worn out mass of the aluminum pin, t represents the time span of the experimental run, V refers to the worn out volume of the pin called the wear volume, H points to the hardness of the sliding pin, F_R is the tangential resistive force between the pin and the disk and is termed as friction force, N is the normal load, and S is the sliding distance on the disk. It is to be noted that the friction coefficient μ is a convenient way to characterize the resistance to relative motion between the surfaces, but it is not a material property nor is it a physical constant. The effect of the polytron additive on different tribological parameters in the experiment and the computed values from the above-mentioned equations are recorded in **Table 8**.

Parameter	Helix base oil (100%)	Helix oil (90%) plus Polytron (10%)
Wear	70 μm	20 μm
Mass wear rate	3.33×10^{-3} mg/min	8.33×10^{-4} mg/min
Volume wear rate	1.28×10^{-3} mm^3/min	6.08×10^{-5} mm^3/min
Coefficient of friction	0.012	0.004
Wear coefficient (k)	—	4.22×10^{-11} m^2/N
Total mass loss	0.7992 mg	0.1992 mg
Total volume loss	0.3079 mm^3	0.01459 mm^3

Table 8.
Computed tribological parameters for the aluminum pin.

3. Results and discussion

The experimentally obtained data and their polynomial fits for the wear behavior of the aluminum metallic pin are displayed in **Figures 3** and **4** for two different configurations in which the experiment was carried out. The adopted test configurations in the experiment were: aluminum pin versus Helix oil-on-steel disk, tagged as AHS configuration, and aluminum pin versus 10% polytron plus 90% Helix oil-on-steel disk, which will be referred to as the APS configuration in the forthcoming discussion.

Figures 3 and **4** show the plot of the wear pattern of the aluminum pin with the passage of time and then the sliding distance on the steel disk of the experiment under consideration. The red line represents wear in the AHS configuration whereas the blue line symbolizes the wear in the APS configuration. It is very much clear from this plot that the polytron additive provides excellent let-up the wear of the tribosystem consisting of an aluminum pin on a steel disk interposed by an oil film of 10% Polytron and 90% Helix. It shows that that the wear in the AHS configuration starts from 70 micron and then stabilizes at approximately 65 micron, but in the APS

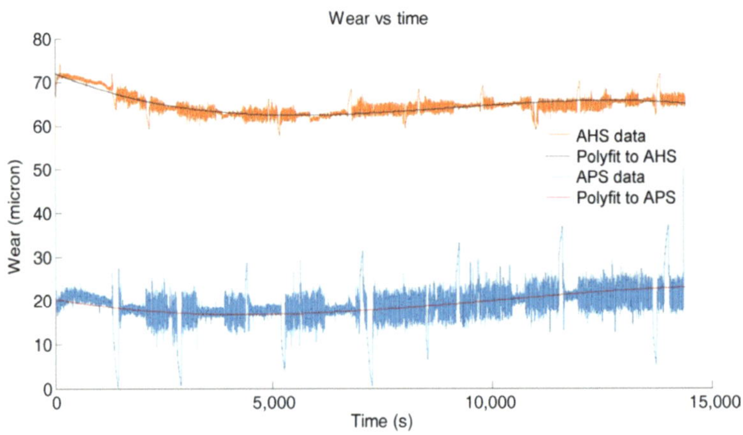

Figure 3.
Graph of the wear of aluminum pin against time in the AHS and APS configurations. The time for the experiment was 240 min.

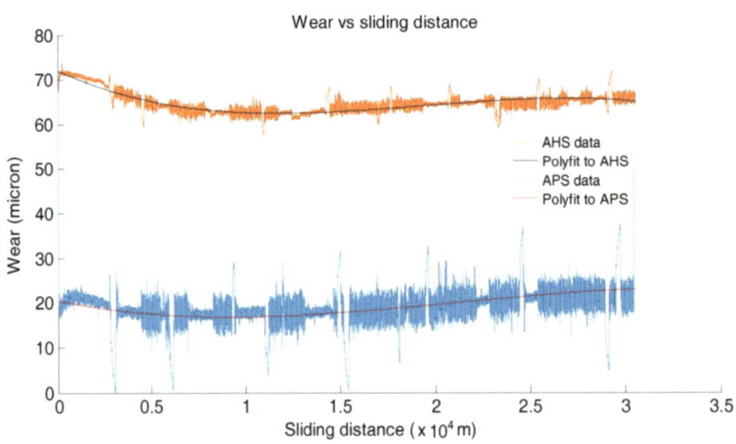

Figure 4.
Graph of the wear of aluminum vs. sliding distance in the AHS and APS configuration. The sliding distance for the experiment was 30.163 km.

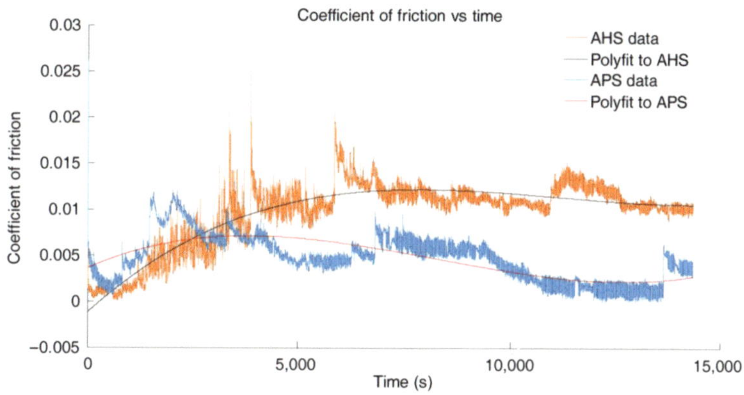

Figure 5.
Evolution of the COF with time in AHS and APS configuration. The experimental time was 240 min.

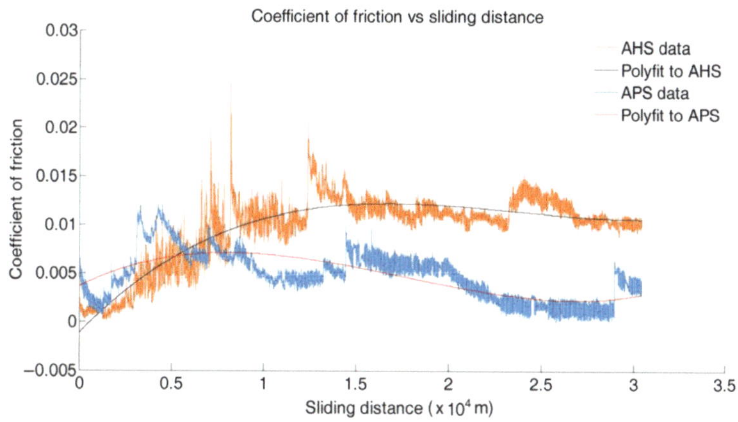

Figure 6.
Evolution of COF with sliding distance in the AHS and APS configuration. The sliding distance for the experiment was 30.161 km.

configuration wear stays at nearly 20 micron. Then, for the same two configurations and under the same experimental conditions, the evolution of the coefficient of friction μ with reference to time span and sliding distance has been plotted as shown in **Figures 5** and **6**.

It is perceivable from the graph of **Figures 5** and **6** that in the AHS format, the initial value of the friction coefficient is almost zero and increases almost linearly to a value of 0.012 in a time span of 100 min of rubbing after which COF stands stable at this very value. The low value of μ in the initial stage of rubbing is probably due to the presence of a layer of foreign material on the disk surface which may be due to some moisture or oxide of the aluminum metal because it readily oxidizes in air. Conversely, in the APS setup, the coefficient of friction starts from a value of 0.005 and then further declines to virtually 0.004. It is recognizable that polytron reduces the wear of the aluminum pin significantly and one can predict that the ratio in the

APS configuration is effectively more than 30% in comparison with AHS configuration. Despite the fact that in our experiment the normal force and sliding distance had very large values in difference with other experimenters, nevertheless the evolved coefficient of friction had negligibly small value when meager 10% polytron was added to 90% helix which in turn endorsed the positive contribution of the polytron in friction minimization. These findings in our tribological

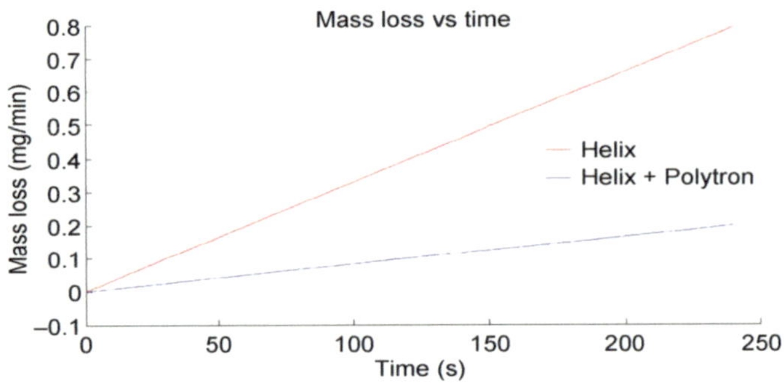

Figure 7.
Graph of the mass loss of aluminum pin vs. time for the AHS and APS configuration. The time of the experiment was 240 min.

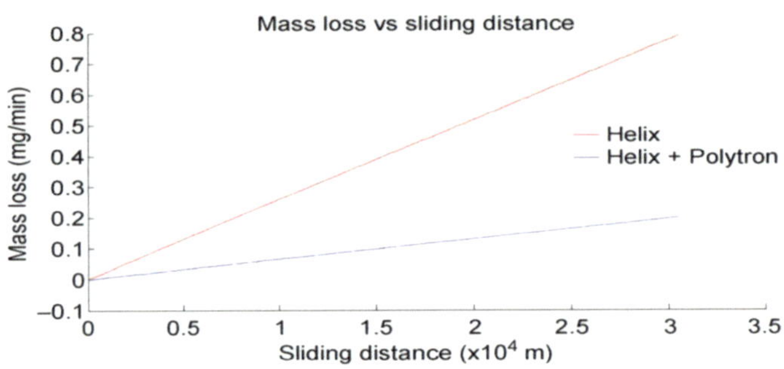

Figure 8.
Experimental graph of the mass loss of aluminum pin vs. sliding distance in the AHS and APS configuration. Sliding distance for the experiment was 30.161 km.

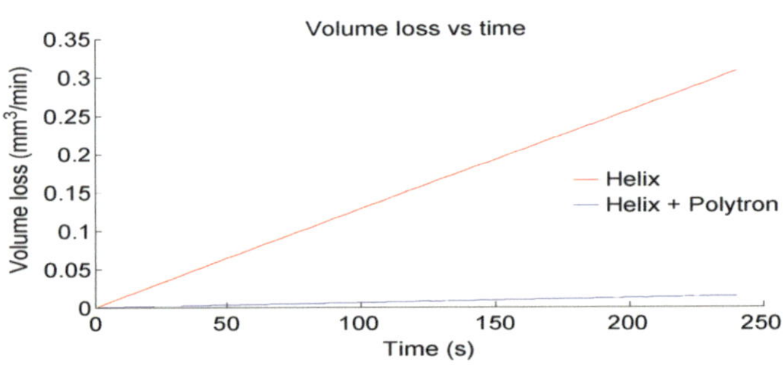

Figure 9.
Graph of the volume loss of aluminum pin vs. time in the AHS and APS configuration. Time for the experiment was 240 min.

experiment with polytron additive in Helix oil are significantly superior in comparison with the findings of other researchers like Nuruzzaman and Chowdhury [57], Bhushan and Kulkarni [58], and Le and Lin [59].

To further clarify the effect of polytron additive, we examined the mass and volume losses of the aluminum pin with regard to time as well as sliding distance and separate graphs were drawn for both the AHS and APS configurations. The comparison plots for the mass losses are shown in **Figures** 7 and **8** while the comparison graphs for the volume losses are illustrated in **Figures 9** and **10**. A deep examination of all the figures reveals that the mass as well as volume loss cannot be controlled with Helix oil alone; rather, it will damage the contact surfaces in a

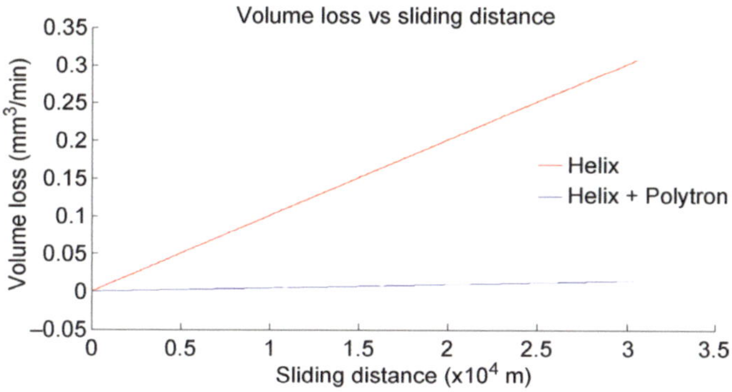

Figure 10.
Graph of the volume loss of aluminum vs. time in the AHS and APS configuration. The sliding distance for the experiment was 30.161 km.

short while, whereas only a scanty addition of 10% of Polytron reduces the mass as well as the volume losses to almost zero level. This is a tremendous change and is visible to the naked eye and directly identifies the supreme antifriction and antiwear capability of the polytron additive. This observation with the addition of polytron additive is fairly divergent with the high wear rate results of researchers like Suarez et al. [60] who studied the popular ZDDP additive in mineral oil stock.

In the same vein, the tribological parameters in our research work are much better than those of Anand et al. [61] who used phosphonium ionic liquid additives in diesel engine lubricants. More to the point, the experimental predictions in our research effort on wear and friction minimization are even far superior to the findings of Chen et al. [62] and Su et al. [63] who used nano-additives in different lubricating media. With an advantage, the calculated values of mass and volume losses of the aluminum pin show that polytron additive attenuates the mass wear rate by an order of magnitude while the volume wear rate of aluminum is alleviated by two orders of magnitude and these outcomes in sequence yield just a nominal value for the wear coefficient as can be noticed from **Table 8**. This outstanding performance identifies that polytron had the capability of permeation into the metal crystal structure of aluminum and subsequent adherence to the metallic surface as an unbreakable surface film that diminished the wear of aluminum surface and consequently curtailed friction between the rubbing surfaces of aluminum and steel.

4. Conclusions

1. The wear of the aluminum metal surface in the Helix base oil was circa 70 μm. The addition of 10% of Polytron additive declined the wear to 20 μm, representing an excess of 2/3 decrement in the wear of the metal.

2. The mass wear rate of the aluminum pin in the Helix base oil was 3.3×10^{-3} mg/min which decreased by an order of magnitude in the Helix plus Polytron mixture, attaining a value of 8.33×10^{-4} mg/min.

3. The mass wear rate of the aluminum pin in the Helix base oil was 1.28×10^{-3} mm^3/min and it decreased by two orders of magnitude in the Helix plus Polytron mixture by assuming a value 6.08×10^{-5} mm^3/min.

4. The value of the coefficient of friction in the Helix oil was estimated at 0.012 which trimmed down to an extremely low value of 0.004 in the combination of 10% polytron additive and 90% Helix oil.

5. Polytron, due to its polar nature, proves to be an effective antiwear additive in the Helix base oil and hence can intrinsically reduce friction by orders of magnitude in mechanical processes and consequently prolong the life span of mechanical parts and, in turn, contribute to considerable fuel and oil economy.

Acknowledgements

The authors are obliged to the management of Universiti Kebangsaan Malaysia for providing laboratory facilities and are especially thankful to the cooperative technical staff of the tribology laboratory. S. M. H. Ahmer and L. S. Jan pay special thanks to Dr. Mohamed Ahmed Siddig of Al-Neelain University, Sudan, and Dr. Siti Fazlili Abdullah of Universiti Tenaga Nasional, Malaysia, for their valuable suggestions.

Conflict of interest

The authors declare that there is no conflict of interest regarding the publication of this paper.

Nomenclature

MTC	metal treatment concentrate
F N	normal force/(load)
F R	resistive force/friction
TCP	tricresylphosphate
ZDDP	zinc dialkyl-diethylthiophosphate
μ	coefficient of friction (COF)

Author details

Syed Mohammad Hassan Ahmer[1], Nusratullah Khan[2], S. Inayat Ali Shah[3] and Lal Said Jan[4]*

1 Department of Physics, Yanbu University College, Yanbu al Sinaiyah, KSA

2 Department of Computer Science, Yanbu University College, Yanbu al Sinaiyah, KSA

3 Islamia College University, Peshawar, Khyber Pakhtunkhwa, Pakistan

4 Department of Physics, Government Postgraduate College, Timergara, Khyber Pakhtunkhwa, Pakistan

*Address all correspondence to: lalsaidjan@gmail.com

References

[1] Bhushan B. Principle and Applications of Tribology. 2nd ed. New York: Wiley; 2013. pp. 344-430. DOI: 10.1002/9781118403020

[2] Neale MJ. The Tribology Handbook. 2nd ed. Oxford: Butterworth-Heinemann; 1995. pp. 300-500. DOI: 10.1115/1.3452677

[3] Blau PJ. Friction Science and Technology: From Concepts to Applications. 2nd ed. Boca Raton: CRC Press; 2009. pp. 225-304. DOI: 978-1-4200-5404-0

[4] Fervel V, Mischler S, Landolt D. Lubricating properties of cotton transfer films studied with a pin-on-disc apparatus. Wear. 2003;**254**:492-500. DOI: 10.1016/S0043-1648(03)00131-5

[5] Hamrock BJ, Schmid SR, Jacobson BO. Fundamentals of Fluid Film Lubrication. 2nd ed. New York: Marcel Dekker; 2004. pp. 22-150. DOI: 0-8247-5371-2

[6] Minami I, Mori S. Antiwear additives for ester oils. Lubrication Science. 2018;**22**:105-121. DOI: 10.1002/jsl.2005.22.2.105

[7] Bhushan B. Handbook of Micro/Nanotribology. 2nd ed. Boca Raton: CRC Press; 1995. pp. 712-742. DOI: 10.1007/978-94-011-5646-2

[8] Kamimura H, Kubo T, Minami I, Mori S. Effect and mechanism of additives for ionic liquids as new lubricants. Tribology International. 2007;**40**:620-625. DOI: 10.1016/j.triboint.2005.11.009

[9] Syahrullail S, Azmi AM, Sapawe N, Amir K. Wear characterization of aluminum lubricated with palm olein at different normal loads. Applied Mechanics and Materials. 2014;**554**:401-405. DOI: 10.4028/www.scientific.net/AMM.554.401

[10] Hisakado T, Hashizume N. Effects of normal load on the friction and wear properties of metal and ceramic against cermet in vacuum. Wear. 2000;**237**:98-106. DOI: 10.1016/S0043-1648(99) 00308-7

[11] Wieleba W. The statistical correlation of the coefficient of friction and wear rate of PTFE composites with steel counterface roughness and hardness. Wear. 2002;**252**:719-729. DOI: 10.1016/S0043-1648(02)00029-7

[12] Ameen HA, Hassan KS, Mubarak EMM. Effect of load, sliding speeds and times on the wear rate for different materials. American Journal of Scientific and Industrial Research. 2011;**2**:99-106. DOI: 10.5251/ajsir.2011.2.99.106

[13] Hisakado T, Miyazaki K, Kameta A, Negishi S. Effects of surface roughness of roll metal pins on their friction and wear characteristics. Wear. 2000;**239**:69-76. DOI: 10.1016/S0043-1648(99)00370

[14] Wang WZ, Chen H, Hu YZ, Wang H. Effect of surface roughness parameters on mixed lubrication characteristics. Tribology International. 2006;**39**:522-527. DOI: 10.1016/j.triboint.2005.03.10.018

[15] Bressan JD, Daros DP, Sokolowski A, Mesquita RA, Barbosa CA. Influence of hardness on the wear resistance of 17-4 PH stainless steel evaluated by the pin-on-disc testing. Journal of Materials Processing Technology. 2008;**205**:353-359. DOI: 10.1016/j.jmatprotec.2007.11.251

[16] Oktay ST, Suh NP. Wear debris formation and agglomeration. Journal of Tribology. 1992;**114**:379-393. DOI: 10.1115/1.2920897

[17] Zhang SC, Pan QL, Yan J, Huang X. Effects of sliding velocity and

normal load on tribological behavior of aged Al-Sn-Cu alloy. Transactions of Nonferrous Metals Society of China. 2016;**26**:1809-1819. DOI: 10.1016/S1003-6326(16)64292-9

[18] Aronov V, Dsouza AF, Kalpakjian S, Shareef I. Experimental investigation of the effect of system rigidity on wear and friction-induced vibrations. Journal of Lubrication Technology. 1983;**105**(2):206-211. DOI: 10.1115/1.3254566

[19] Aronov V, Dsouza AF, Kalpakjian S, Shareef I. Interactions among friction, wear, and system stiffness—Part 1: Effect of normal load and system stiffness. Journal of Tribology. 1984;**106**:54-58. DOI: 10.1115/1.3254567

[20] Berger EJ, Krousgrill CM, Sadeghi F. Stability of sliding in a system excited by a rough moving surface. Journal of Tribology. 1997;**119**(4):672-680. DOI: 10.1115/1.2833868

[21] Carcel AC, Palomares D, Rodilla E, Puig MP. Evaluation of vegetable oils as prelube oils for stamping. Materials & Design. 2005;**26**:587-593. DOI: 10.1016/j.matdes.2004.08.010

[22] Husnawan M, Saifullah MG, Masjuki HH. Development of friction force model for mineral oil base stock containing palm olein and antiwear additive. Tribology International. 2007;**40**:74-81. DOI: 10.1016/j.triboint.2006.02.062

[23] Kalin M, Vižintin J. A comparison of the tribological behaviour of steel/steel, steel/DLC and DLC/DLC contacts when lubricated with mineral and biodegradable oils. Wear. 2006;**261**:22-31. DOI: 10.1016/j.wear.2005.09.006

[24] Menezes PL, Kishore, Kailas SV. Studies on friction and transfer layer: The role of surface texture. Tribology Letters. 2006;**24**:265-273. DOI: 10.1080/10402000902825754

[25] Lin JW, Bryant MD. Reductions in wear rate of carbon samples sliding against wavy copper surfaces. Journal of Tribology. 1996;**118**:116-124. DOI: 10.1115/1.2837065

[26] Petlyuk AM, Adams RJ. Oxidation stability and tribological behavior of vegetable oil hydraulic fluids. Tribology Transactions. 2004;**47**:182-187. DOI: 10.1080/05698190490431849

[27] Du L, Xu B, Dong S, Yang H, Tu W. Study of tribological characteristics and wear mechanism of nano-particle strengthened nickel-based composite coatings under abrasive contaminant lub-rication. Wear. 2004;**257**:1058-1063. DOI: 10.1016/j.wear.2004.07.003

[28] Chowdhury MA, Helali MM. The effect of amplitude of vibration on the coefficient of friction. Tribology International. 2008;**41**:307-314. DOI: 10.1016/j.triboint.2007.08.005

[29] Ustunyagiz E, Sulaiman MH, Christiansen P, Bay N. A study on DLC tool coating for deep drawing and ironing of stainless steel. Key Engineering Materials. 2018;**767**:181-188. DOI: 10.4028/www.scientific.net/KEM.767.181

[30] Ligier JL, Noel B. Friction reduction and reliability for engines bearings. Lubricants. 2015;**3**:569-596. DOI: 10.3390/lubricants3030569

[31] Tung SC, McMillan ML. Automotive tribology overview of current advances and challenges for the future. Tribology International. 2004;**37**:517-536. DOI: 10.1016/j.triboint.2004.01.013

[32] Bermúdez MD, Jiménez AE, Sanes J, Carrión FJ. Ionic liquids as advanced lubricant fluids. Molecules. 2009;**14**:2888-2908. DOI: 10.3390/molecules14082888

[33] Qu J, Luo H, Chi M, Ma C, Blau PJ, Dai S. Comparison of an oil-miscible

ionic liquid and ZDP as a lubricant anti-wear additive. Tribology International. 2014;**71**:88-97. DOI: 10.1016/j.triboint.2013.11.010

[34] Otero I, López ER, Reichelt M, Villanueva M, Salgado J, Fernandez J. Ionic liquids based on phosphonium cations as neat lubricants or lubricant additives for a steel/steel contact. Applied Materials & Interfaces. 2014;**6**:13115-13128. DOI: 10.1021/am502980m

[35] Totolin V, Minami I, Gabler C, Brenner J, Dörr N. Lubrication mechanism of phosphonium phosphate ionic liquid additive in alkylborane–imidazole complexes. Tribology Letters. 2014;**53**:421-432. DOI: 10.1007/s11249-013-0281-0

[36] González R, Battez AH, Viesca J, Higuera-Garrido A, Fernández-González A. Lubrication of DLC coatings with two tris(pentafluoroethyl) trifluorophosphate anion-based ionic liquids. Tribology Transactions. 2013;**56**:887-895. DOI: 10.1080/10402004.2013.810319

[37] Viesca J, García A, Battez AH, González R, Monge R, Fernández-González A. FAP anion ionic liquids used in the lubrication of a steel–steel contact. Tribology Letters. 2013;**52**:431-443. DOI: 10.1007/s11249-013-0226-7

[38] Battez AH, González R, Viesca JL, Blanco D, Asedegbega E, Osorio A. Tribological behavior of two imidazolium ionic liquids as lubricant additives for steel/steel contacts. Wear. 2009;**266**:1224-1228. DOI: 10.1016/j.wear.2009.03.043

[39] Blanco D, González R, Hernández BA, Viesca J, Fernández GA. Use of ethyl-dimethyl-2 methoxyethylammonium tris (pentafluoroethyl) trifluorophosphate as base oil additive in the lubrication of TiN PVD coating. Tribology International. 2011;**44**:645-650. DOI: 10.1016/j.triboint.2011.01.004

[40] Yu B, Bansal DG, Qu J, Sun X, Luo H, Dai S. Oil-miscible and noncorrosive phosphonium-based ionic liquids as candidate lubricant additives. Wear. 2012;**289**:58-64. DOI: 10.1016/j.wear.2012.04.015

[41] Saka N, Liou MJ, Suh NP. The role of tribology in electrical contact phenomena. Wear. 1984;**100**:77-105. DOI: 10.1016/0043-1648(84)90007-3

[42] Bakunin VN, Suslov AY, Kuzmina GN, Vedeneeva LM, Parenago OP, Migdal CA, et al. Surface-capped molybdenum sulphide nanoparticles—A novel type of lubricant additive. Lubrication Science. 2004;**16**:207-214. DOI: 10.1002/ls.3010160302

[43] Liu YB, Lim SC, Ray S, Rohatgi PK. Friction and wear of aluminum-graphite composites: The smearing process of graphite during sliding. Wear. 1992;**159**:201-205. DOI: 10.1016/0043-16489(92)90303-p

[44] Zhang L, Chen L, Wan H, Chen J, Zhou H. Synthesis and tribological properties of stearic acid-modified anatase (TiO_2) nanoparticles. Tribology Letters. 2011;**41**:409-416. DOI: 10.1007/s11249-010-9724-z

[45] Zhang BS, Xu BS, Xu Y, Gao F, Shi PJ, Wu YX. Copper nanoparticles effect on the tribological properties of hydro-silicate powders as lubricant additive for steel-steel contacts. Tribology International. 2011;**44**:878-886. DOI: 10.1016/j.triboint.2011.03.002

[46] Cumings J, Zettl A. Low-friction nano-scale linear bearing realized from multiwall carbon nanotubes. Science. 2000;**289**:602-604. DOI: 10.1126/science.289.5479.602

[47] Stachowiak GW, Batchelor AW. Engineering Tribology. 4th ed. Amsterdam: Elsevier; 2014. pp. 200-255. DOI: 10.1016/C2011-0-075-4

[48] Quaroni L, Chumanov G. Preparation of polymer-coated functionalized silver nanoparticles. Journal of the American Chemical Society. 1999;**121**:10642-10643. DOI: 10.1021/ja992088q

[49] Mandal T, Fleming MS, Walt DR. Preparation of polymer-coated gold nanoparticles by surface-confined living radical polymerization at ambient temperature. Nano Letters. 2002;**2**:3-7. DOI: 10.1021/nl015582c

[50] Benabdallah HS, Wei JJ. Effects of lubricants on the friction and wear properties of PTFE and POM. Journal of Tribology. 2005;**127**:766-775. DOI: 10.1115/1.2005276

[51] Zhang Z, Liu W, Xue Q. Study on lubricating mechanisms of $La(OH)_3$ nano-cluster modified by compound containing nitrogen in liquid paraffin. Wear. 1998;**218**:139-144. DOI: 10.1016/S0043-1648(98)00225-7

[52] Hu ZS, Dong JX. Study on antiwear and reducing friction additive of nanometer titanium oxide. Wear. 1998;**216**:92-96. DOI: 10.1016/S0043-1648(97)00252-4

[53] Ahmer SMH, Jan LS, Siddig MA, Abdullah SF. Experimental results of the tribology of aluminum measured with a pin-on-disk tribometer: Testing configuration and additive effects. Friction. 2016;**4**:124-134. DOI: 10.1007/s40544-016-0109-7

[54] John EH. Aluminum: Properties and Physical Metallurgy. Ohio: ASM International; 1984. pp. 11-20. DOI: 10.1361/appm1984p025

[55] Ahmer SMH, Bahrudin MS, Abdullah SF, Jan LSA. Quantitative analysis of the wear of aluminum in the presence of polytron additive in the helix lubricant. In: Proceedings of the National Graduate Conference (NatGrad'12); 8-10 November 2012; UNITEN: Malaysia

[56] Odabas D. Effects of load and speed on wear rate of abrasive wear for 2014 Al alloy. Materials Science and Engineering. 2018;**295**:1-12. DOI: 10.1088/1757-899X/295/1/012008

[57] Nuruzzaman DM, Chowdhury MA. Effect of load and sliding velocity on friction coefficient of aluminum sliding against different pin materials. American Journal of Materials Science. 2012;**2**:26-31. DOI: 10.5923/j.materials.20120201.05

[58] Bhushan B, Kulkarni AV. Effect of normal load on micro-scale friction measurements. Thin Solid Films. 1996;**278**:49-56. DOI: 10.1016/0040-6090(95)08138-0

[59] Le VNA, Lin JW. Tribological properties of aluminum nanoparticles as additives in an aqueous glycerol solution. Applied Science. 2017;**7**:80-95. DOI: 10.3390/app7010080

[60] Suarez AN, Grahn M, Pasaribu R, Larsson R. The influence of base oil polarity on the tribological performance of zinc dialkyl-dithiophosphate ZDDP additives. Tribology International. 2010;**43**:2268-2278. DOI: 10.1016/j.triboint.2010.07.016

[61] Anand M, Hadfield M, Viesca JL, Thomas B, Hernández Battez A, Austen S. Ionic liquids as tribological performance improving additive for in-service and used fully-formulated diesel engine lubricants. Wear. 2015;**334**:67-74. DOI: 10.1016/j.wear.2015.01.055

[62] Chen S, Liu WM, Yu LG. Preparation of DDP-coated PbS nanoparticles and investigation of

the antiwear ability of the prepared nanoparticles as additive in liquid paraffin. Wear. 1998;**218**:153-158. DOI: 10.1016/S0043-1648(98)00220-8

[63] Su Y, Gong L, Chen D. An investigation on tribological properties and lubrication mechanism of graphite nanoparticles as vegetable based oil additive. Journal of Nanomaterials. 2015;1-7. DOI: 10.1155/2015/276753

Chapter 3

Structural-Energy Interpretation of the Friction

Sergey Fedorov

Abstract

The structural-energy model of elastic-plastic deformation is considered as the main mechanism of transformation and dissipation of energy under friction. The equations of friction energy balance are proposed. The energy interpretation of the coefficient of friction is given. A structural-energy diagram of the friction surfaces is proposed. The energy regularities of evolution of tribological contact (elementary tribosystem) are discussed. The idea of the smallest structural element of dissipative friction structures (mechanical (nano) quantum) is discussed. Mechanical quantum is dynamic oscillator of dissipative friction structure. The nano-quantum model of the surfaces damping is proposed. Calculations for some Hertzian heavily loaded contacts of real tribosystems are proposed.

Keywords: energy balance, contact evolution, adaptation, dissipation, nanostructure, wear standard

1. Introduction

Modern tribology considers elastic-plastic deformation of friction surfaces as the main mechanism of transformation and dissipation of energy during friction.

The modern view of plastic deformation offers ergodynamics of deformable solids [1–3]. Ergodynamics of deformable solids is a synthesis to the problem of deformation most general laws of thermodynamics for non-reversable processes, molecular kinetics and dislocation theory in their mutual, dialectical tie on the basis of a most general law of nature—the law of energy conservation at its transformations.

The macroscopic phenomenon of plastic deformation, damage and destruction of a solid element is considered as a set of a huge number of microscopic elementary acts of atomic-molecular rearrangements, causing the generation (reproduction) by sources, movement, interaction and destruction of various kinds of elementary defects on the drains. Each defect is a carrier of excess potential energy and on its formation is spent strictly defined work of external forces.

From the thermodynamic point of view, the whole variety of mechanisms and structural levels of plastic flow can be divided into two most characteristic groups—adaptive and dissipative types. The first group should include the mechanism of nucleation and accumulation in the local volumes of various kinds of elementary defects and damages of the structure. The second group includes elementary acts of atomic-molecular rearrangements associated with the movement and destruction of various defects on the drains, that is, controlling the dynamic return.

Such structural-energy interpretation of plastic deformation (friction of contact volumes) determines kinetic and competitive regularities of the process [1–3].

If you apply the basic concepts of plastic deformation of solids this theory for the analysis of the process of friction, it is possible to consider the method of triboergodynamics [4].

The major distinction of triboergodynamics from general ergodynamics of deformed solids is "scale factor" which exhibits itself in existence of critical friction volume. This volume determines the limit friction parameters and separate, in essence, the surface deformation from the traditional volume deformation.

In the most general case the triboergodynamics should be seen as structural-energy interpretation of the friction process. In the framework of triboergodynamics the process of friction is considered as an evolutionary phenomenon of the contact friction (rubbing surfaces).

2. Short fundamentals of ergodynamic of deformed solids

2.1 Structural model of the material

The deformable body is considered as an open, multicomponent, essentially inhomogeneous and nonequilibrium system, representing a hierarchy of statistically uniformly distributed over the volume of metastable structural elements (defects and damages) of various (from macro- to micro-) levels. Some of these structural elements are virtual sources and sinks of elementary defects (vacancies, dislocations, etc.), others—obstacles to their movement.

The main parameters characterizing the structural state of the material are [2, 3]: γ_σ is the coefficient of overstress on interatomic bonds, characterizing the uneven distribution of external stresses σ on interatomic bonds $\sigma^0 (\gamma_\sigma = \sigma^0/\sigma \geq 1)$; u_e is the density of latent (free) energy of defects and damages; v is the coefficient of unevenness of the distribution of latent energy in volume, representing the ratio between the density of latent energy in the local volume u_e^0 to the average value $u_e (\nu = u_e^0/u_e)$. The complex structural parameter $k = \gamma_\sigma / v^{0,5} = \sigma_* / S_*$ characterizes the relationship between the theoretical σ_* and real S_* strength of a solid.

2.2 Physical model and structural-energy interpretation of the process

Macroscopic phenomena of plastic deformation and scattered destruction of the body element are a cooperation of a huge number of microscopic elementary acts of atomic and molecular rearrangements in the field of external (thermal, mechanical, electrical, etc.) forces activated by thermal energy fluctuations. The whole variety of mechanisms and structural levels of the process from the thermodynamic point of view is divided into two most characteristic groups—adaptive and dissipative (relaxation) type, which differ in physical nature and kinetic laws. The first group includes elementary acts that control the origin and accumulation of elementary defects in the deformable body (damageability). The integral characteristic of intensity of the specified processes is specific (referred to unit of volume) power of pumping of excess (latent) energy $\dot{u}_e$

$$\dot{u}_e = \frac{du_e}{dt} = A \sinh\left[\left(\alpha\sigma_i^2 - \nu\, u_e\right)/2kT\right]. \tag{1}$$

The second group includes mechanisms and elementary acts that control relaxation (dissipative) processes of plastic deformation. The integral characteristic of these processes is the specific power of the thermal effect $\dot{q}$ of plastic deformation

$$\dot{q} = \frac{dq}{dt} = B \sinh\left[\left(\alpha\sigma_i^2 + \nu\, u_e\right)/2kT\right]. \tag{2}$$

here, A and B are the kinetic coefficient

$$A = \frac{2kT}{hV_0}\sum_1^n U_i'(\sigma_0, T)\exp\left[-\frac{U'(\sigma_0, T)_i}{kT}\right], \tag{3}$$

$$B = \frac{2kT}{hV_0}\sum_1^n U_i''(\sigma_0, T)\exp\left[-\frac{U''(\sigma_0, T)_i}{kT}\right], \tag{4}$$

$$U_i'(\sigma_0, T) = U_{0i}' + \Delta U'(T) \pm \beta\sigma_0^2,\ U_i''(\sigma_0, T) = U_{0i}'' + \Delta U''(T) \pm \beta\sigma_0^2, \tag{5}$$

$$\alpha = \frac{\gamma_\sigma^2 V_0}{6G}, \beta = \frac{\gamma_\sigma^2 V_0}{2K}, \tag{6}$$

where U'_{0i}, U''_{0i} is the activation energy of the formation and diffusion of the ith defect, respectively; σ_0, σ_i is the hydrostatic stress and stress intensity; V_0 is the atomic volume; k is the Boltzmann constant; h is the Planck constant; T is the absolute temperature; G, K is the shear and bulk elasticity modules.

2.3 Thermodynamic analysis of interrelation between deformation and fracture

From the thermodynamic point of view, the process of plastic deformation and destruction is characterized by the competition of two opposite, interrelated and simultaneously occurring trends in the body element—the growth of the latent energy density u_e of various defects and damages arising and accumulating in the material due to the work of external forces ω_p, and its reduction (release) due to relaxation processes occurring inside the deformable body element; in this case, the first trend is associated with the deformation hardening and material damage, the second-with the dynamic return and dissipation of strain energy, causing the thermal effect q of plastic deformation.

A significant part of the dissipation energy q is not retained in the deformable element of the body, passes through it as if in transit and dissipates in the environment due to heat exchange $\vec{q}$. Only a small part of the dissipation energy q accumulates in the deformable element of the body in the form of a thermal component of the internal energy $\Delta u_T = q - \vec{q}$ increasing its temperature (self-heating effect).

In accordance with the law of conservation and transformation of energy

$$\omega_p = \Delta u_e + q \quad \text{and} \quad \dot{\omega}_p = \dot{u}_e + \dot{q}. \tag{7}$$

In the mechanics of a deformable solid, irreversible work ω_p and the power $\dot{\omega}_p$ of deformations are associated with the stress-strain state of the body element by the relation

$$d\omega_p = \sigma_i d\varepsilon_i^p,\ \dot{\omega}_p = \sigma_i \dot{\varepsilon}_i^p. \tag{8}$$

where $\dot{\varepsilon}_i^p$ is the rate of irreversible deformation.

Joint consideration Eqs. (7) and (8) allows to establish a unique relationship between the stress-strain and thermodynamic states of the body element

$$\dot{\varepsilon}_i^p = \frac{\dot{\omega}_p}{\sigma_t} = \frac{1}{\sigma_i}(\dot{u}_e + \dot{q}) = \dot{\varepsilon}_i^e + \dot{\varepsilon}_i^q. \qquad (9)$$

Therefore, from the thermodynamic point of view, the total values of the work ω_p and irreversible deformation ε_i^p and the rates of their change $(\dot{\omega}_p, \dot{\varepsilon}_i^p)$ can be represented as the sum of two terms associated, respectively, with the deformation hardening and damage $(\dot{\varepsilon}_i^e = \dot{u}_e/\sigma_i)$ and dynamic return $(\dot{\varepsilon}_i^q = \dot{q}/\sigma_i)$ controlling quasi-viscous flow of the body element.

This important conclusion is of fundamental importance in the analysis of the relationship between the processes of deformation and destruction of the body element. For damage and destruction of the body element is responsible only part of the plastic (irreversible) deformation ε_i^e controlled by microscopic processes associated with deformation hardening and accumulation of latent energy of defects and damages. A significant part of the irreversible deformation ε_i^q controlled by relaxation (dissipative) processes does not affect the damage and destruction of the body element, but only causes its quasi-viscous flow (stationary creep). The relationship between the work and the degree of irreversible deformation and their components varies within a very wide range and depends on the structure of the material and the conditions of its deformation [1].

2.4 Thermodynamic condition of local fracture

The parameter of damage (scattered destruction) is taken as the density of the internal energy u accumulated in the deformable volumes, determined by the sum of two components: potential (latent) u_e and kinetic (thermal) u_T that is

$$\Delta u = \Delta u_e + \Delta u_T, \dot{u} = \dot{u}_e + \dot{u}_T. \qquad (10)$$

This energy is associated with the accumulation in the deformable element of the body of static (Δu_e) and dynamic (Δu_T) damages and distortions of the crystal lattice, therefore, is dangerous, responsible for the scattered destruction (damage). The element of the body is considered to be destroyed if at least one local micro-volume responsible for the destruction, the density of internal energy reaches a critical (limit) value u_* corresponding to the loss of crystal lattice stability "in a large." This point corresponds to the appearance in the local micro-volume of a crack of critical size (according to Griffiths-Orovan-Irvin) and a sharp localization of the process at the mouth (top) of the crack. The thermodynamic condition of local fracture is written as

$$u(\overline{r}_*, t_*) = u(\overline{r}_*, 0) + \int_0^{t_*} \dot{u}(\overline{r}_*, t)dt = u_* = const. \qquad (11)$$

here, $u\ (\overline{r}_*, 0)$ is the density of internal energy in the local micro-volume of the material in the initial (before deformation $t = 0$) state; $\dot{u}\ (\overline{r}_*, t)$is the specific power of internal energy sources in the local volume responsible for the destruction; $\overline{r}_*$ is the parameter characterizing the coordinates (x_*, y_*, z_*) of the local volume responsible for the fracture.

2.5 Thermodynamic criterion of fracture

In accordance with the structural-energy analogy of the process of mechanical destruction and melting of metals and alloys [5] and theoretical and experimental studies [1, 6], the critical value of the internal energy density u_* in the local macro-volume of the material responsible for destruction coincides well with the known thermodynamic characteristic of the material ΔH_S is the enthalpy of melting, that is.,

$$u_* = \Delta H_s = \int_0^{T_s} c_p dT + L_s. \quad (12)$$

here, T_S is the melting temperature; c_p is the heat capacity; L_S is the latent heat of melting.

3. Triboergodynamic's method

3.1 Friction

3.1.1 Initial or zero axiom of friction

The present day analysis of sum total of modern friction investigations may be presented in the form of three theses (others are also possible) of essential property which are shared by many research workers as undoubt proof as to the most characteristic properties of generalized friction model:

1. Friction is the phenomenon of resistance to the relative movement (movement) of surfaces, localized at the points of contact tangent to them;

2. Friction is the process of converting (transforming) the energy of external mechanical motion into other types of energy, and mainly into thermal energy;

3. Friction is a process of elastic-plastic deformation and fracture localized in thin surface layers of friction pair materials.

These three axioms may be regarded as initial friction axioms and called "zero" friction axioms as the starting-point of whence it is possible to develop logical analysis of generalized engineering property for friction process.

In the capacity of axiomatic method of friction investigation of initial friction axioms [4] mentioned above the author thinks it expedient to use the method of ergodynamics of deformable solids [1–3] which are at present may be taken as axiomatic, that is, method which may be trusted owing to the theoretical, experimental and practical substantiation.

3.1.2 Balanced and unitary attributes of friction

Taking into consideration that basic attribute of any system is a balance attribute then tribosystem framework should be determined by the framework of obeying, for example, energy balance friction. Then it follows that basic equation for tribosystem is an energy balance equation characterizing movement within friction system in a generalized and quantitative way. Constituent parts of this balance must determine basic quantitative regulations of energy transformations (and movement) within the system.

Thus, tribosystem in the most generalized sense is quantitatively characterized by the energy balance equation. Most generalized quantitative regulatities of tribosystem behavior (states) are determined by magnitudes relations among constituencies of friction energy balance. These conditions may also be taken as friction axioms. In accordance with that it is possible to show justice of entropy balance equation and so of information and etc.

Taking into consideration the fact that the most characteristic magnitude of the most global balance principle is unit (whole), then, consequently, the basic parameters of tribosystem (friction), expressed as indexes of relations among balance constituents must also have criterion (limit) magnitudes equal to unit.

3.1.3 Common energy analysis of friction process

In the most general case the work of friction process W_F is summed up from the work of elastic W_F^{elast} and plastic W_F^{plast} deformation and wear (failure) of contact volumes (**Figure 2**) and work for overcoming forces of viscous friction and failure of lubricant material W_{lub}:

$$W_F = W_F^{elast} + W_F^{plast} + W_{\text{lubr}}, \tag{13}$$

For particular case of friction without lubrication ($W_{\text{lub}} \cong 0$) and in the conditions of stationary (developed) friction, when the work of elastic deformation may be neglected due to their insignificance, friction work W_F will be determined mainly by the work of plastic deformation of surfaces (contact volumes) of shaft W_{F1}^{plast} and of bearing W_{F2}^{plast}:

$$W_F = W_F^{plast} = W_{F1}^{plast} + W_{F2}^{plast}. \tag{14}$$

3.2 Structural-energy interpretation of friction process

It is known friction is characterized a product of frictional forces F by friction distance ℓ, that is., the work ω_f, expended on overcoming frictional forces

$$\omega_f = F\ell, \tag{15}$$

$$\omega_f = \Delta u_e + q, \tag{16}$$

$$\dot{\omega}_f = \dot{u}_e + \dot{q}. \tag{17}$$

here, $\dot{\omega}_f = d\omega_f/dt$ is a power of friction dissipation of energy; $\dot{u}_e = du_e/dt$ is the rate of storing latent energy in deformed (contact) volumes; $\dot{q} = dq/dt$ the power of thermal effect of plastic deformation (friction).

Since the contact volumes of both materials that make up the friction pair are deformed by friction (see **Figure 2**), Eqs. (16) and (17) should be written as

$$\omega_f = \Delta u_{e1} + \Delta u_{e2} + q_1 + q_2, \tag{18}$$

$$\dot{\omega}_f = \dot{u}_{e1} + \dot{u}_{e2} + \dot{q}_1 + \dot{q}_2. \tag{19}$$

These equations show, that from thermodynamic point of view, the work ω_f of friction forces, (friction power $\dot{\omega}_f$) is related to plastic deformation of the contact volumes. The work ω_f may be divided conventionally into two specific parts.

The first part of the friction work is related to the change in the deformable (contact) volumes of materials of latent (potential) energy Δu_{e1} and Δu_{e2}. It is the energy of various elementary defects and damages arising and accumulating in deformable volumes. This energy is a unique and integral characteristic of submicro- and microstructural changes that occur in plastically deformable volumes of materials [1, 2, 7]. It is a measure of deformation hardening and damage of materials.

The second part of the friction work ω_f is related to the processes of dynamic return, accompanied by the release of latent energy and the thermal effect q_1, q_2 of friction. This energy is associated with the movement and destruction of various elementary defects of opposite signs, their exit to the surface, healing reversible submicroscopic discontinuities, etc.

The relations between the components of the energy balance of the friction process Δu_{e1} and Δu_{e2}, as well as q_1 and q_2 vary widely and are determined by the physical and chemical properties of the materials that make up the friction pair, their structure and the conditions of the friction process.

In the most general case, Eqs. (18) and (19) should be presented (**Figure 2**) taking into account the real (not unit) sizes of the tribocontacts.

$$W_f = \Delta U_e + Q = \Delta U_{e_1} + \Delta U_{e_2} + Q_1 + Q_2, \tag{20}$$

$$\dot{W}_f = \dot{U}_e + \dot{Q} = \dot{U}_{e_1} + \dot{U}_{e_2} + \dot{Q}_1 + \dot{Q}_2, \tag{21}$$

where $\Delta U_e = V_f \Delta u_e$; $\dot{U}_e = V_f \dot{u}_e$; V_f is contact (deformed) volume of the materials of the friction pair.

Solving Eqs. (20) and (21) with respect to the friction force F, we obtain generalized equations for the friction force

$$F_l = \frac{\Delta U_{e_1} + \Delta U_{e_2}}{l} + \frac{Q_1 + Q_2}{l}. \tag{22}$$

$$F_v = \frac{\dot{U}_{e_1} + \dot{U}_{e_2}}{v} + \frac{\dot{Q}_1 + \dot{Q}_2}{v}, \tag{23}$$

where l and v are the friction path and the slip velocity.

Dividing both parts of the Eqs. (22) and (23) by the normal force N, we present generalized equations for the coefficient of friction

$$\mu_l = \frac{\Delta U_{e_1} + \Delta U_{e_2}}{Nl} + \frac{Q_1 + Q_2}{Nl}, \tag{24}$$

$$\mu_v = \frac{\dot{U}_{e_1} + \dot{U}_{e_2}}{Nv} + \frac{\dot{Q}_1 + \dot{Q}_2}{Nv}. \tag{25}$$

Thus, friction is generally described by the equation of energy balance and from the thermodynamic point of view [1–4] it is a competitive process of two (mentioned above) opposite, interrelated and simultaneously occurring in the deformable contacts trends. According to the energy balance scheme (**Figure 1**) for plastic deformation and fracture [1] presented above (relationships $\Delta u = \Delta u_e + \Delta u_T$ and $q = \Delta u_T + \vec{q}$), equations [8] for friction work W_f, frictional force F and friction coefficient μ (without lubrication) has view

$$W_f = \Delta U_e + Q = \Delta U_{e_1} + \Delta U_{e_2} + \Delta U_{T_1} + \Delta U_{T_2} + \vec{Q}_1 + \vec{Q}_2, \tag{26}$$

$$\dot{W}_f = \dot{U}_e + \dot{Q} = \dot{U}_{e_1} + \dot{U}_{e_2} + \dot{U}_{T_1} + \dot{U}_{T_2} + \dot{\vec{Q}}_1 + \dot{\vec{Q}}_2, \tag{27}$$

$$F_l = \frac{\Delta U_e}{l} + \frac{Q}{l} = \frac{\Delta U_{e_1} + \Delta U_{e_2}}{l} + \frac{Q_1 + Q_2}{l}, \tag{28}$$

$$F_v = \frac{\dot{U}_{e_1} + \dot{U}_{e_2}}{v} + \frac{\dot{Q}_1 + \dot{Q}_2}{v} = F_{mechanical} + F_{molecular}, \tag{29}$$

$$\mu_l = \frac{\Delta U_{e_1} + \Delta U_{e_2}}{Nl} + \frac{Q_1 + Q_2}{Nl} = \mu_{adapt} + \mu_{dis} = \mu_{adapt} + \mu_{T(dis)} + \mu_{\vec{Q}(dis)}, \tag{30}$$

$$\mu_v = \frac{\dot{U}_{e_1} + \dot{U}_{e_2}}{Nv} + \frac{\dot{Q}_1 + \dot{Q}_2}{Nv} = \mu_{deformation} + \mu_{adhesion}, \tag{31}$$

where $\Delta U_e = V_f \Delta u_e$; $Q = V_f q$; $\vec{Q} = V_f \vec{q}$; $\dot{U}_e = V_f \dot{u}_e$; $\dot{u}_e = \mathrm{d}\, u_e / \mathrm{d}\, t$ is the rate of latent energy density change in the contact volumes; V_f is a deformable volume of friction; μ is the coefficient of friction; μ_{adapt} is the adaptive coefficient of friction; $\mu_{T(dis)}$ and $\mu_{\vec{Q}(dis)}$ are the static and dynamic components of dissipative coefficient of friction; ΔU_T is the thermal component of internal energy; N is the normal load; l is the friction distance; v is the sliding velocity. The latent energy density Δu_e is an integral parameter of tribostate and damageability (failure (Δu_e^*)) of solids [1].

Thus, viewed thermodynamically, the work done by friction forces W_f (the friction power $\dot{W}_f$), the friction force F and the friction coefficient μ may be

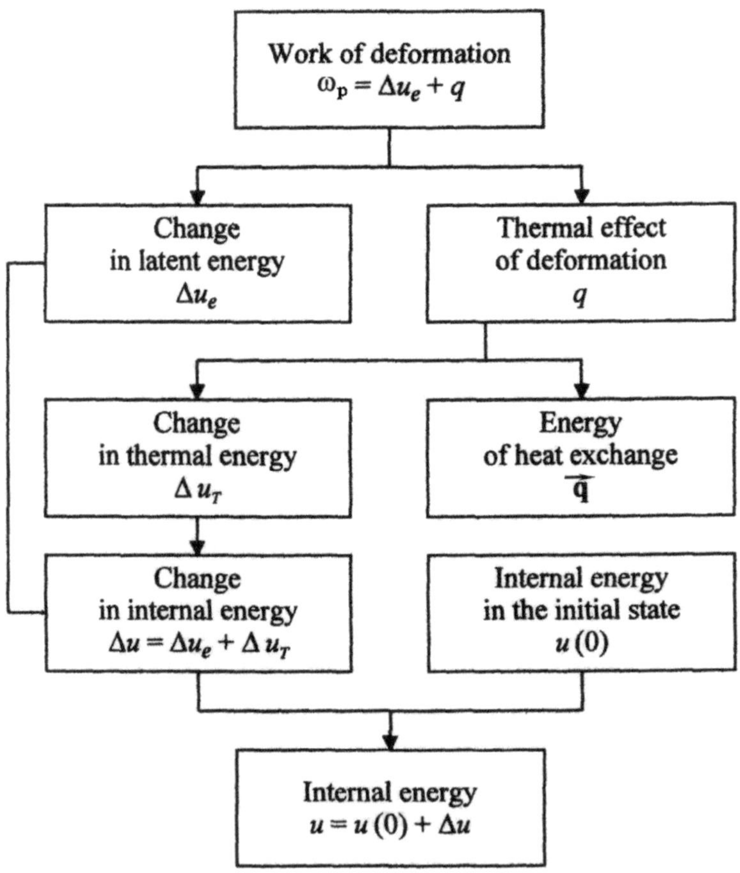

Figure 1.
Scheme of the energy balance for the plastic deformation (friction) of a solid [1–3].

classified conventionally into two specific components with different kinetic behavior [3, 9]. The first component is associated with microscopic mechanisms of adaptive type and relates to the change of latent (potential) energy ($\Delta\, u_{e_1}$, $\Delta\, u_{e_2}$) of various elementary defects and damages that are generated and accumulate in the deformable volumes of materials friction pair (**Figure 1**). This energy is a unique and integral characteristic of the submicro- and microstructural transformations that occur in plastically strained materials [1–3, 9]. It is a measure of deformation hardening and damage of materials. The second component is associated with microscopic mechanisms of dissipative type and related to dynamic recovery processes in which latent energy is released and heat effect of friction (q_1, q_2) take place. This energy is associated with the movement and destruction of various elementary defects of opposite signs, their exit to the surface, healing reversible submicroscopic discontinuities, etc. The ratios of the components $\Delta\, u_{e_1}$ and $\Delta\, u_{e_2}$ as well as q_1, q_2 of the balance vary over a wide range, depending on the physical, chemical, and structural properties of the materials that comprise the friction couple and the friction process conditions [8].

Thus, the thermodynamic analysis of the plastic deformation and fracture of the solid volume at friction allows us to obtain generalized (two-term) dependences for the friction force F and the friction coefficient μ, which corresponds to the modern concepts of the dual nature of friction [10, 11]. It is a molecular mechanical Eq. (29) and deformation-adhesion Eq. (31) theories of friction. But, more correctly it is necessary to speak about adaptive-dissipative nature (model) of friction Eq. (30).

As follows from the equations of the energy balance of friction Eqs. (26) and (27), the whole variety of manifestations of friction and wear can be conditionally reduced to at least two fundamentally different states. The first condition determines all types of damageability and wear, the second-the so-called condition of “wearlessness” [7].

The state of damageability and wear is characterized by the components of energy balance Eqs. (26) and (27), which are responsible for accumulation of internal energy $\Delta\, u = \Delta\, u_{e1} + \Delta\, u_{e2} + \Delta\, u_{T1} + \Delta\, u_{T2}$ in deformed volumes, that is, the process is irreversible [4, 8]. The “wearlessness” state is characterized by the components of the energy balance Eqs. (26) and (27), which are responsible for the dynamic dissipation (reversibility) of strain energy into elastic and structural dissipated energy $\vec{q} = \vec{q}_1 + \vec{q}_2$ of friction contact [4, 8].

In its turn, the first state may be classified depending on the relation between potential $\Delta\, u_e$ and kinetic $\Delta\, u_T$ components of internal energy. It is subdivided conventionally into mechanical damage and wear (due to so-called structure activation) and thermal damage and wear (due to thermal activation). For instance, let the thermal component of internal energy $\Delta\, u_T$ be equal to zero ($\Delta\, u_T = 0$) and the internal energy variation at damage and wear be defined only by variation of the potential $\Delta\, u_e(\Delta\, u = \Delta u_e)$ component. Then, the mechanical damage and wear with brittle fracture of surfaces take place. On the contrary, if we have $\Delta\, u_e = 0$ ($\Delta\, u = \Delta\, u_T$), then the thermal damage and wear with ductile fracture of surfaces take place. All the intermediate values of the components are associated with quasi-brittle or quasi-ductile fracture of solids [4, 8].

In the most general case, taking into account a fundamental tribology’s notion of the “third body” [10], the energy balance at dry friction Eq. (20) should be written as

$$W_f = \Delta U_{e1} + \Delta U_{e2} + \Delta U_{e3} + Q_1 + Q_2 + Q_3. \tag{32}$$

In the special case, where the friction is localized into volume of the “third body” (**Figure 2**) Eq. (32) develops into

$$W_f = \Delta U_{e3} + \vec{Q}_3. \qquad (33)$$

here, $\Delta U_{e3} = V_3 \Delta u_{e3}$.

3.3 Energy interpretation of the friction coefficient by Amonton (Leonardo da Vinci)

According to the main conclusion of the thermodynamic theory of strength [1], as a structural parameter should not take the entire value of the accumulated plastic deformation, but only its part associated with the deformation hardening, which is uniquely and integrally determined by the density of the potential component of the internal energy (i.e., the density $\Delta\, u_e$ of the so-called latent energy) of various defects and damages accumulated in the plastically deformable volumes of the material. With this in mind, if we neglect the heat effect Q of friction, one will infer from the thermodynamic analysis of friction of Eqs. (24) and (25) that the Amonton (Leonardo da Vinci) friction coefficient is

$$\mu = \frac{\Delta U_e}{\mu^* Nl} = \frac{F}{N}; F = \frac{\Delta U_e}{l}; Q \cong 0, \mu^* = 1. \qquad (34)$$

Consequently, the coefficient of friction has a very deep physical sense. On the one hand, it is the parameter which generally characterizes the resistance of relative displacement (movement) of surfaces, for it reflects the portion of energy, which “is done by friction away” as accumulated latent energy ΔU_e, by relation to parameter of external forces work $\mu^* Nl$ (energy of external relative movement) [12]. On the other hand, it is the generalized characteristic of damage, for it is defined of the latent energy density $\Delta\, u_e$ as integral characteristic of the structure defectiveness measure, because this energy is the generalized parameter of damage. Here too, coefficient of friction generally reflects the structural order (disorder) of deforming contact volume, since the parameter $\Delta U_e = \Delta u_e V_f$ is defined of the energy of defects and damages of different types, that are accumulated into contact volumes V_f solids [12].

Thus, the coefficient of friction is a true and generalized parameter of the state of the tribosystem. It follows a very important conclusion that the analysis of the regularities of the evolution of the states of tribosystems is, first of all, the analysis

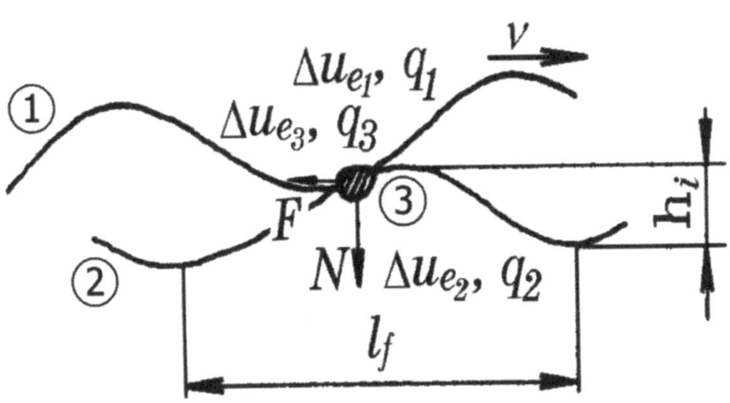

Figure 2.
Conditional scheme of friction contact [4].

of the laws of change of the accumulated latent energy of deformation by the contacting volumes of the solid, that is, change of Amontons coefficient of friction [12].

3.4 Generalized experimental friction curves

The dependences obtained for the friction coefficient μ are in agreement with experimental curves $\mu = \mu(N, v)$ (**Figures 3–5**). Analyzing various experimental friction curves using the Eqs. (20)–(31) of friction energy balance, it was concluded

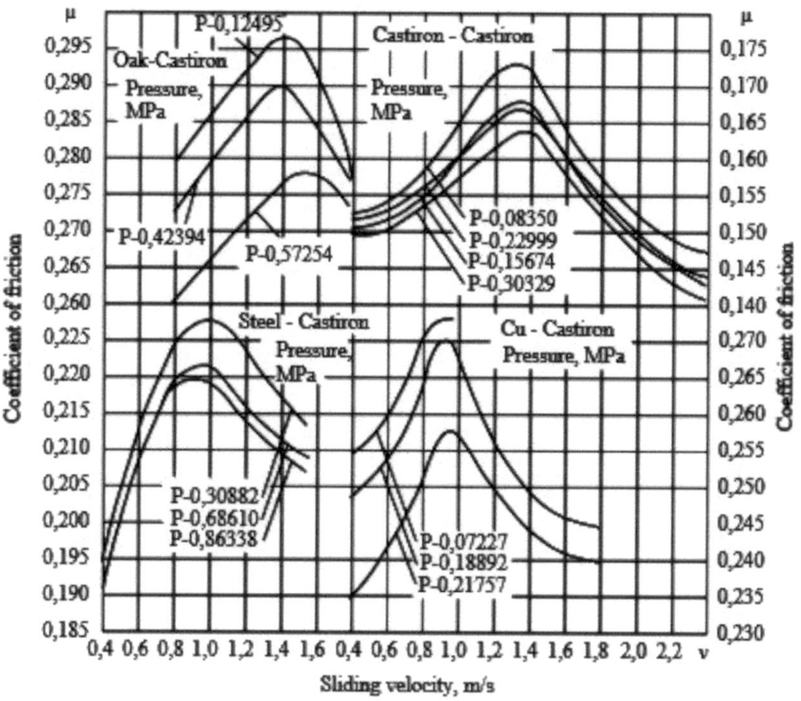

Figure 3.
Experimental results of Conti [13].

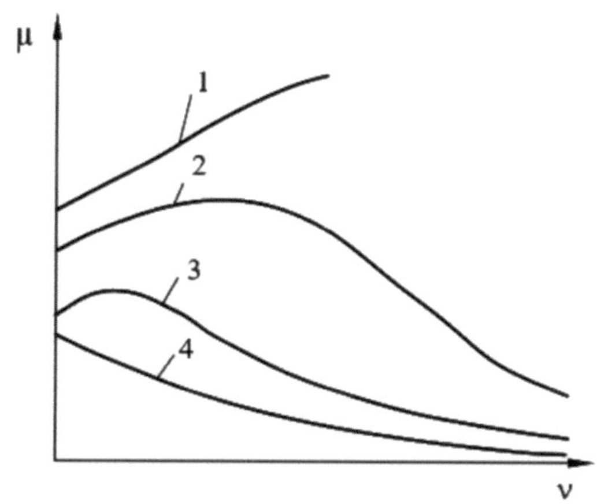

Figure 4.
Generalized friction experiments in I.V. Kragelsky's interpretation [10]: sliding velocity (load: 1—small; 2 and 3—medium; 4—considerable).

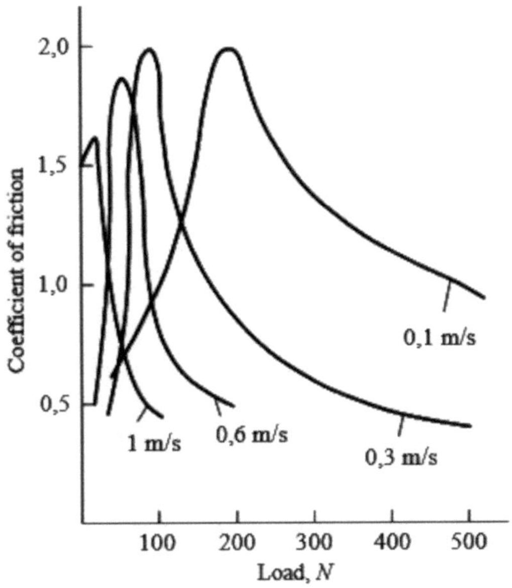

Figure 5.
Experimental results of Watanabe for a pair of friction—nylon 6—steel [14].

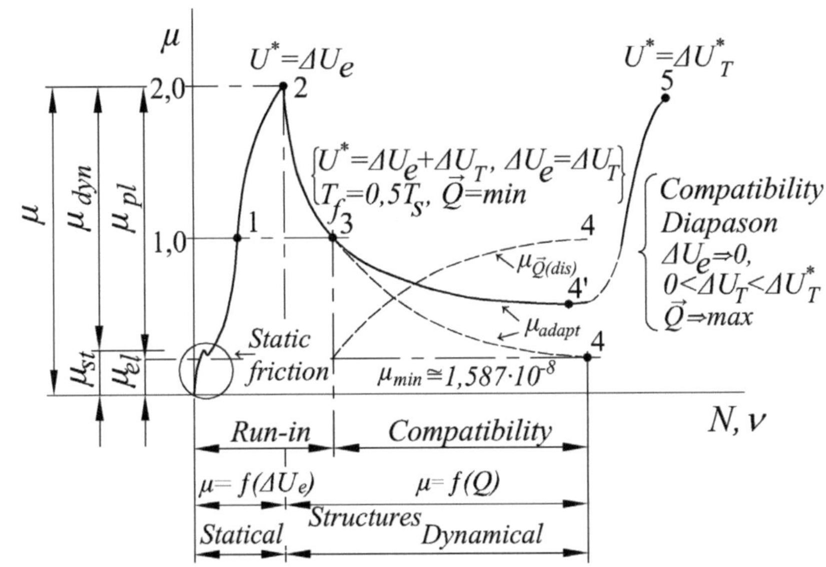

Figure 6.
Structural-energy diagram of the evolution of friction surfaces [4, 16–18].

[4] that the experimental friction curves (**Figures 3–5**) of the type a $\mu = \mu(N, v)$ are generalized experimental friction curves and reflect the general (for all materials and friction pairs) laws of evolution (changes in the friction coefficient) of tribosystems.

3.5 Structural-energy regularities of rubbing surfaces evolution

An analysis of modern experimental data using Eqs. (20)–(31) has shown that the experimental friction curves of type $\mu = \mu(N, v)$ are the generalized experimental friction curves that reflect the evolution (the change in the friction coefficient) of tribosystem.

We propose an energetic interpretation of the experimental friction curves $\mu = \mu(N, v)$ (**Figure 6**). According to our concept [4, 15, 16], the ascending portion of the friction coefficient curve μ is mainly controlled by processes associated with the accumulation of latent energy ΔU_e in various structural defects and damages. Here the increase in μ is due to the increasing density of latent (potential) energy $\Delta\ u_e$ and the increasing adaptive friction volume V_f. The descending portion of the friction curve is mainly controlled by processes associated with the release and dissipation of energy $Q = \Delta U_T + \vec{Q}$. Here the decrease in μ is due to the decrease in latent energy density within the friction volume V_f or (which is virtually the same) to the decrease of the adaptive friction volume V_{adapt} $(u_e = u_e^*)$ and to the increase of the dissipative volume V_{dis} $\left(\vec{q}^{\,*} = u_e^*\right)$.

The evolution of the tribosystem is presented in the form of a diagram (**Figure 6**) and has an adaptive-dissipative character Eqs. (29)–(34) and reflects the competitive (dialectical) nature of friction. The evolutionary curve has a number of fundamental points (1–5) of transition states of the tribosystem, which are strictly subject to the balance principle of friction. Between these points there are the most characteristic areas of behavior of the tribosystem. These areas reflect the most general properties of nonlinear dynamics of friction evolution.

So, in **Figure 6** you can see the following conditionally marked points and areas: 0-1—the area of static friction and strain hardening; 1—the point of the limit strain hardening; 1-2—excess energy pumping area; 2—point of adhesion (seizure) and transition of external friction into internal (point of critical instability); area of formation of dissipative structures (formation of temperature fluctuation in the friction volume); 3—the point of minimum compatibility (maximum frictionness); 1-2-3—area of self-organization; 3-4—compatibility area; 4—point of wearlessness (abnormally low friction); 5—thermal adhesion point.

The ideal evolution of the friction contact is symmetric. The friction process begins and ends in areas of elastic behavior. Between them is the plastic maximum (super activated state) as a condition of self-organization and adaptation. In the most general case, the regularities of evolution (adaptation) of tribosystems can be represented as two-stage (**Figure 6**). At the first stage (0–2) of the evolution of the friction contact, it tends to form a critical volume V_f^* of friction (point 2). This is the smallest volume of friction that has accumulated the maximum potential energy of structure defects. This is an elementary tribosystem, that is., an elementary and self-sufficient energy transformer. In the first stage, the latent energy density Δu_e increases to a limit value Δu_e^* within the critical friction volume V_f^*.

The volume of friction V_f^* is constant in the second stage of evolution. At this stage, contact is evolutionarily developed due to structural transformation. At this stage, a wide spectrum of compatible friction structures (**Figure 6**) can be formed depending on the nature of the environment. The second stage (2–4) can be considered as a structural transformation of the critical friction volume V_f^* (elementary tribosystem) conditionally to the adaptive V_{adapt} and dissipative V_{dis} friction volumes (**Figure 7**). The end point (point 4) of this stage of evolution is characterized by the complete transformation of the critical adaptive friction volume V_{adapt}^* into the dissipative V_{dis}^* one.

The above volumes mentioned characterize different regularities of energy conversion of external mechanical motion at friction. Adaptive volume V_{adapt}

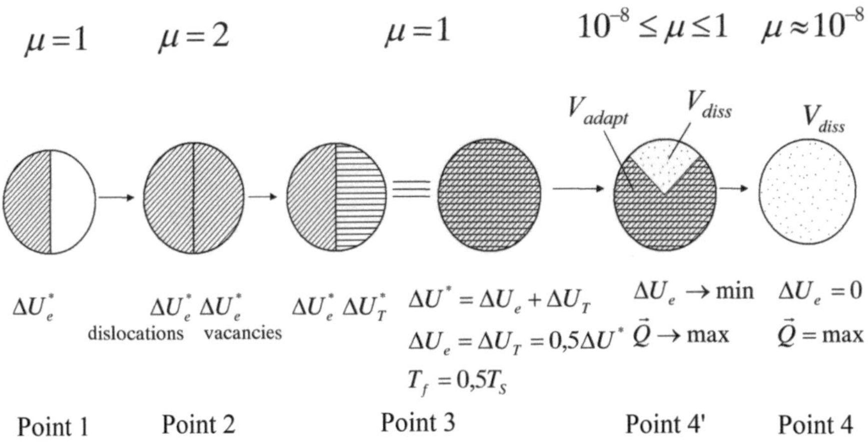

Figure 7.
*The model of the energy evolution of the contact friction volume at points 1–5 of the diagram (**Figure 6**) [4].*

is associated with irreversible absorption of strain energy. In this volume, there is an accumulation of latent deformation energy Δu_e and centers of destruction are born. The dissipative volume V_{dis} is able to reversibly transform (dissipate) the energy of the outer movements. In this volume, there is no accumulation of latent deformation energy due to the flow of reversible elastic-viscoplastic deformation.

Theoretical and calculated estimates [4, 16, 18] have shown that the dissipative friction volume performs a reversible elastic transformation of the energy of external mechanical motion with a density $\vec{q}^{\,*}$ equal to the critical density u_e^* of the latent energy.

The culmination of the evolution of the tribosystem is its final and limiting state of point 4—the state of abnormally low friction and wearlessness (maximum efficient).

A schematic evolution of the contact volume of friction in diagram's points 1–5 is presented in **Figure 7**.

Calculations show [4] that with the ideal evolution of the tribosystem, the adaptive (Amonton) coefficient of friction μ_{adapt} at point 2 of the diagram drops sharply, reaching at point 4 the elastic coefficient of friction μ_{elast}. For point 4 of compatibility area 3-4, an equation of energy balance Eq. (30) should be put in the following way:

$$\mu_{adapt} = \mu^* - \mu_{dis} = 1 - \mu_{dis} = \mu_{plast} = 0 = \mu_{elast}; \mu^* = 1, 0. \tag{35}$$

Thus, we have at point 4 of the ideal evolution of the contact friction volume the condition of perfectly elastic-viscous-plastic deformation. This actually shows the Eq. (35), that is, the coefficient of friction of Amonton μ_{adapt}, being, in fact, a plastic coefficient of friction μ_{plast} has a minimum value equal to zero. Consequently, plastic friction becomes elastic with the coefficient μ_{elast} of friction. This means that the plastic deformation of the contact friction volume is realized with the maximum dynamic dissipation ($\vec{Q}$= max) of the accumulated latent energy. Therefore, the amount of accumulated energy at point 4 is zero ($\Delta U_e = 0$). This fact proves the ideal state with the full evolution of the contact volume. From the physical point of view, this state can be explained by the complete dissipating of the energy ΔU_e^* accumulated at point 2, along the newly formed structures of point 4 in the form of elastic energy of interaction between them (energy $\vec{Q}^{\,*}$ of dynamic dissipation). Here $\mu_{dis} = 1, 0$. The structural elements themselves are defectlessness—$\mu_{adapt} = 0$, and friction is elastic—$\mu = \mu_{elast}$.

It is shown [4] that the value of the minimum adaptive friction volume $V_{adapt}^{\min}$ corresponding to the zero value of the plastic friction component μ_{adapt} is not zero, but is equal to the size of a certain minimum structural element of the deformable solid.

3.6 The idea of a mechanical (nano) quantum of dissipative friction structures

The result of ideal elementary tribosystem (contact) evolution is forming of unique nanostructure—a mechanical (nano) quantum. Strict ideas about mechanical quantum obtained [4, 18] considering for point 4 of the friction evolution diagram the equation of a quasi-ideal solid:

$$\vec{Q}^{*} = \vec{S}_Q T = \mu_{dis}^{*} N l_f = V_f^{*} u_e^{*} = V_f^{*} \vec{q}_{*}. \tag{36}$$

This equation is a special case of the solution of the equations of energy balance of friction Eq. (29), at $\mu_{adapt} = 0$ and $\mu_{dis} = 1 = \mu_{dis}^{*}$. Here $\vec{S}_Q$is the inertial entropy of compatible friction volume; T is the characteristic temperature of compatible contact friction volume; l_f is the linear dimension of elementary contact.

Accordingly, under the conditions of maximum compatibility (point 4), when the tribosystem implements a complete evolutionary cycle of adaptation with the formation of the most perfect, dissipative structure, its (structure) behavior is subject to the equation of state of a quasi-ideal solid, that is, it should be assumed that the interactions between the elements of this structure are minimized—the state of ideal elasticity in dynamics. Eq. (28), taking into account the Planck-Boltzmann formula $S = k \ln W$ and the real number of atomic oscillators N_f in the volume of the elementary tribosystem (contact) V_f^{*}, is given to the form explaining the regularities of friction in terms of the evolution of systems:

$$\mu_{\text{diss}} = \frac{\vec{S}_Q}{N l_f} = \frac{kTN_f \ln W}{N l_f}, \tag{37}$$

$$\mu_{adapt} = 1 - \mu_{diss} = 1 - \frac{kTN_f \ln W}{N l_f} = 1 - \frac{\vec{S}_Q\, T}{N l_f} = \frac{S_U T}{N l_f}, \tag{38}$$

where k is the Boltzmann constant; W is the probability of state; S_U is the configuration entropy of friction (contact) volume.

The tribosystem always tends to some optimal state characterized, that is, to the most probable state $W' = N_f \ln W$ for the given friction conditions.

The analysis and solution of these equations [4, 16–18] allows to show the principle of the constancy of the magnitude of the probability (the state's parameter (order)) W of the tribological system) for the entire range of compatible friction, namely $\ln W = 3$, and $W = e^3 = 20, 08553696....$

The number of thermodynamic state probability W equal to 20, 08553696... was interpreted [4, 12–15] as the smallest number of linear, atomic oscillators in one of the three directions of the minimum adaptive friction volume $V_{adapt}^{\min}$ corresponding to the state of almost absolute elastic friction—abnormally low friction (safe deformation threshold). Accordingly, the number of atomic oscillators in this volume is $V_Q = (e^3)^3 = (20, 08553695...)^3 = 8103, 083969....$

It is the universal size (volume) of mechanical quantum [4, 7, 16–18].

On the other hand, taking the meaning of Boltzmann entropy S, we obtain a universal friction constant $R_f = kN_f$ [4, 16–18], which in physical sense characterizes the "energy size" of the elementary tribosystem (TS) containing under ideal conditions the same number of atomic oscillators N_f (mechanical quanta N_Q):

$$R_f = k \cdot N_f = k \cdot W^3 \cdot N_Q = R_{MQ} \cdot N_Q, \frac{J}{grade \cdot TS}, \tag{39}$$

$$R_{MQ} = k \cdot W^3, \frac{J}{grade \cdot MQ}, \tag{40}$$

where R_{MQ} is the universal constant of deformation at friction.

As follows from the calculations [4], the size of the minimum adaptive friction volume $V_{adapt}^{\min}$ coincides in its magnitude with the size of the submicroscopic zone at the mouth of the crack, which for metals is equal $(4...9) \cdot 10^{-6}$ mm, that is, with the size of the critical volume responsible for the fracture. Thus, the size of the minimum adaptive friction volume $V_{adapt}^{\min} = V_{elast}$ can be represented as the size of some mechanical "quantum".

This mechanical quantum is the minimum number of atoms capable of providing a configuration of their distribution (structure), which has the property of reversibly absorb and dissipate (return) the energy of external mechanical motion (action). It also represents the smallest structural formation under plastic deformation and is formed during the transition of the tribosystem (deformable volume) through the extremely activated (critical) state (**Figure 6**) due to the development of self-organizational processes of adaptation of the tribosystem. The mutual rotational-oscillatory motion of these mechanical quanta relative to each other inside the elementary tribosystem (contact) determines the state of the most perfect dissipative structure of friction. Actually, this state is described by the equation of state of a quasi-ideal solid Eq. (36), the state when the interaction between the elements of the structure (mechanical quanta) is minimized-the state of the ideal elasticity of the quasi-viscous flow. The calculated coefficient of friction between the quanta is approximately 10^{-8} [4, 16–18].

The conclusion that the mechanical quantum is the smallest structural formation under plastic deformation (friction) is confirmed by the calculation. If we compare the values of the elastic modules E to the atomic (true) elasticity E_r, we obtain values equal to 60, where the number $60 = 3W$ can be interpreted as a characteristic of the volume elasticity of one mechanical quantum—the minimum adaptive friction volume $V_{adapt}^{\min}$. Calculation of the parameter $W \cong 20 = E/3E_r$ for different metals and steels gives an average value of 20, 77 (**Table 1**); $\Delta H_S = 3E_r$ is the enthalpy of melting.

It is concluded [4, 16–18] that for all materials, under the conditions of the ideal evolution of the tribosystem, the number of atoms N_f (mechanical quanta (MQ)) in the volume of one elementary tribosystem (TS) is constant. Thus, we can talk about the amount of matter equal in mass to one elementary tribosystem and to one mechanical quantum.

3.7 The synergy of the tribosystem and the optimality states

Mechanical quantum is a dynamic oscillator of dissipative friction structures. The ideal, quasi-elastic state of contact at its full evolution is the effect of the most complete energy dissipation of external mechanical motion on the

Metals and steels	$E \times 10^{-3}$, MPa	$(u_e^*)\Delta H_S \times 10^{-3}$, MJ/m^3	$E/3E_r$
Cr	235.4	8.5	27.69
Mg	44.4	1.9	23.37
Ag	79.0	3.7	21.35
Au	78.7	4.0	19.67
Co	200.1	10.6	18.88
Fe	211.4	9.9	21.35
Ta	184.4	10.6	17.39
Ti	105.9	6.7	15.8
Nb	104.0	9.2	11.3
Zr	95.6	5.7	16.77
Mo	316.9	12.0	26.4
W	392.4	14.4	27.25
Ni	201.1	9.4	21.39
Iron	210.9	10.1	20.88
20	200.1	9.5	21.06
1Kh13	206.0	8.9	23.14
3Kh13	218.8	9.2	23.78
Kh18N9T	199.1	9.4	21.19
Kh18M9	199.1	9.6	20.74
30Kh	214.1	10.2	20.99
30N3	207.5	10.3	20.11
40	209.4	9.7	21.58
30G2	207.2	10.0	20.72
30KhGN3	208.0	10.2	20.4
G13	204.0	10.0	20.4
50S2G	196.2	10.3	19.05
U8	198.0	10.3	19.22
U12	198.0	10.4	19.04

$\Delta H_S = 3E_r$, $E/3E_r = 20.77$.

Table 1.
Parameter estimation for different metals and steels [4].

newly formed (on the mechanism of self-organization in the vicinity of the critical state) structural elements-mechanical quanta (dynamic oscillators), which implement the most complete rotational-oscillatory behavior relative to each other in the volume of the elementary tribosystem. At the same time, the resistance to their relative interaction is minimal-elastic and corresponds to the elasticity of ideal atomic (thermodynamically equilibrium) interactions at the level of electronic shells.

The universal constants of the mechanical quantum and the elementary tribosystem (material point) determine both the quantum model of surface damping:

$$\mu_{dis} = \frac{3R_{MQ}Tn_i}{Nl_f} = \frac{U_{1Q}n_i}{U_{1Q}n_*} = \frac{n_i}{n_*} = 1 - \mu_{adapt}; \quad \mu_{adapt} = 1 - \frac{n_i}{n_*} = \frac{n_{dest}}{n_*}, \tag{41}$$

taking into account the quanta of destruction n_{dest} (irreversible component of the process) and the quanta of damping n_i (reversible, elastic component (fatigue number)), and the probabilistic model of the evolution of the tribosystem to the most ordered state:

$$\mu_{adapt} = 1 - \mu_{dis} = 1 - \frac{R_f T \ln W_i}{Nl_f} = 1 - \frac{\ln W_i}{\ln W_*}. \tag{42}$$

where $3R_{MQ}T = U_{1Q}$ is the energy of one mechanical quantum; W_i and W_* is the current and limit probabilities of states of compatible tribosystems.

According to the model of quantum damping of surfaces under friction in the conditions of the most complete evolution (adaptation) of the elementary tribosystem, all mechanical quanta except one, elastically and reversibly transform the energy of external action (mechanical motion). One mechanical quantum of radiation ($\cong$ 8103 atoms)—there is a minimum loss (the essence of wearlessness (the ideal damping properties) or the standard of wear).

The linear size of a mechanical quantum is equal to the diameter of a spherical ideal crystal with atomic roughness [4, 7]:

$$D_{MQ} = 2 \cdot W \cdot \overline{d}_a \cdot (3/4 \cdot \pi)^{1/3} = 7,177 nm. \tag{43}$$

here, $\overline{d}_a$ is the average atomic diameter, for metals; $W = e^3$ is the mechanical quantum state parameter [4].

The mechanical quantum (**Figure 8**) itself should be considered as an elementary nanostructure of a metal solid.

Figure 8.
Model of an ideal crystal of elementary nanostructure of friction contact (8103 atomic cubical cells) [4, 16–18].

Calculations have shown [4, 8] the number N_Q of such mechanical “quanta” (subtribosystems) within the elementary tribosystem’s volume $V_f^* = V_{dis}^*$ to be $0.63 \cdot 10^8$ which is close to the safe number n_* of fatigue cycles.

Therefore, the smaller the coefficient of friction μ_{adapt} (the greater the coefficient μ_{dis}) of the tribosystem, the higher its fatigue endurance (durability), as a greater number of mechanical quanta involved in the process of damping (elastic return) of the energy of the external mechanical motion (impact), and consequently the smaller the number of quanta associated with the fracture (accumulation of latent energy of defects and damage of the limit value). In the limit, the tribosystem is characterized by the effect of “wearlessness” (abnormally low friction), corresponding to the state of almost complete thermodynamic reversibility of the friction (deformation) process. Here, all mechanical quanta, with the exception of one, reversible elastic transform (damp out) the energy of external mechanical movement. By analogy with classical quantum theory, we can say that in this case the system (tribosystem) is in the ground state (here, as if all mechanical quanta are directed against the field)—tribosystem cannot give energy to any other system (environment) simply because it (tribosystem) and does not accumulate energy in this state. In this case, the tribosystem is in almost perfect balance with the environment.

The principle of mechanical quantum determines nanoquantum levels of all friction parameters of compatible (optimal) tribosystems and other.

4. The model for the evaluation of wear of compatible friction

The model [4] of the moving critical (equilibrium) friction volume (**Figure 9**) is considered for the analysis of wear problems.

Here, the instantaneous value of the friction work $\dot{W}_{f_i}$ is connected with the friction work $\dot{W}_f$ per unit time, taking into account the uniform distribution of contacts (micro-shocks) in the longitudinal n_v and transverse $n_н$ directions of the friction surface:

$$\dot{W}_f = \dot{W}_{f_i} n = \dot{W}_{f_i} n_н n_v = \dot{W}_{f_i} \frac{H}{l_f} \cdot \frac{L_v}{l_f}. \quad (44)$$

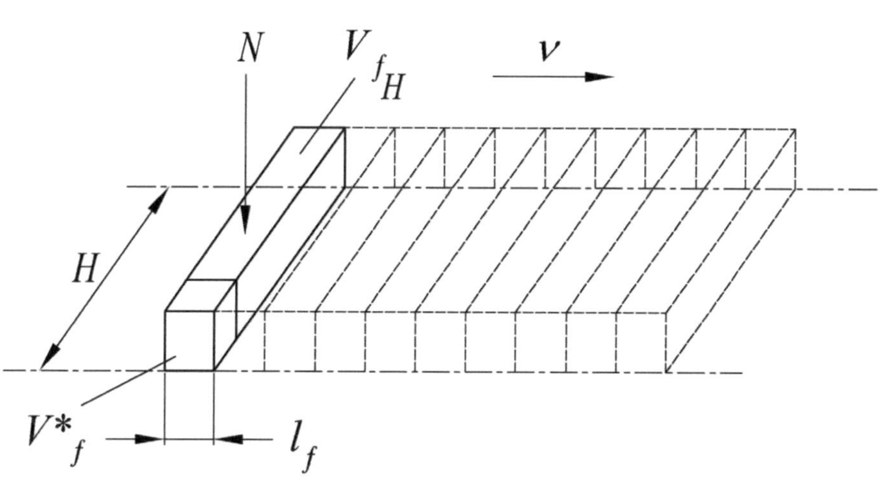

Figure 9.
Scheme to the calculation of wear parameters of friction [4].

Accordingly, we have a number of ratios for power, force and coefficient of friction

$$\dot{W}_f^* = \dot{W}_{f_i}^* \cdot n = V_f^* \cdot \Delta u_e^* \cdot n = V_{f_H}^* \cdot \Delta u_e^* \cdot n_v, \tag{45}$$

$$F^* = \frac{\dot{W}_f^*}{v} = \frac{V_f^* \cdot \Delta u_e^* \cdot n}{l_f \cdot n_v} = \frac{V_{f_H}^* \cdot \Delta u_e^*}{l_f}, \tag{46}$$

$$\mu^* = \frac{\dot{W}_T^*}{N^* \cdot v} = \frac{V_f^* \cdot \Delta u_e^* \cdot n}{N^* \cdot v} = \frac{V_{f_H}^* \cdot \Delta u_e^*}{N^* \cdot v}. \tag{47}$$

Here $V_{f_H}^* = V_f^* \cdot n_H$ (**Figure 9**); W_T^{i*} is the instantaneous (contact) the value of the friction work; $n = \frac{V_f^t}{V_f} = n_v \cdot n_H$ is the ratio of the volume of friction V_f^t deformable per unit time t to the instantaneous volume of friction V_f; n_v, n_H is the number of micro-shocks in the sliding direction of the sample per unit time and in the transverse direction.

Eq. (47), performed to the form of the $\mu^* = \frac{h_r \cdot \Delta u_e^*}{p_r \cdot l_f} = \frac{h_a \cdot \Delta u_e^*}{p_a \cdot B}$, represents the basic equation of wear for compatible friction region:

$$\tau_r = \frac{h_r}{l_f} \Delta u_e^* = I_r \Delta u_e^*, \tag{48}$$

$$\tau_a = \frac{h_a}{B} \Delta u_e^* = I_a \Delta u_e^*. \tag{49}$$

here, I_r, I_a is the linear wear rate, related to the real and nominal areas of contact; B, H is the sample sizes in the slide and the longitudinal directions.

5. Nano quantum models of the maximum capacity for work of the tribosystem

5.1 The principle of calculating the wear of gears

All parameters of compatible (optimal) friction should be in nanoquant levels, which are commensurate with the parameters of one mechanical quantum—standard of wear.

Operation of all heavily loaded tribosystems should be considered from the standpoint of the ideal evolution of tribosystems. This is a perfect condition contact friction is the true indicator of the state of the tribosystem for practical examples of tribology. This is the standard of maximum efficiency of the tribosystem—abnormally low friction and wearlessness.

A typical example of wear (destruction) of real tribosystems on the model of mechanical quantum is the work of gears (for example, reducers) and systems of wheel-rail and other, in which the elementary particle of wear (pitting) is wear equal to one mechanical quantum. Imagine the engagement of a pair of teeth involute profile on the field of the length of the active line of engagement (**Figure 10**) as the model of smooth surfaces with uniformly distributed equilibrium roughnesses after run-in (elementary tribosystems, which are analogues of the material point of mechanics). Engagement of a pair of teeth corresponds to the theoretical principle of running two cylinders under the conditions of Hertz elastic-plastic contact. The materials of the teeth work at the limit of the fatigue threshold, which corresponds to the minimum loss (pitting) of the contact volume (elementary tribosystem) in the form of a single mechanical quantum.

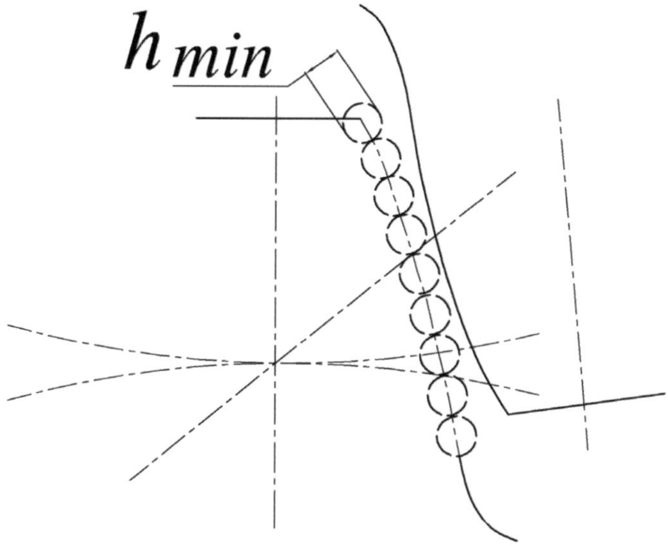

Figure 10.
Model of active tooth surface of a gear wheel with equilibrium roughness of spherical shape [4].

When working gear engagement, for each revolution of the wheel (gear), each roughness (material point) of the active surface of the tooth is loaded once, with a minimum loss (wear) in one mechanical (nano) quantum. Since the critical volume of friction (elementary tribosystem) contains $0.63 \cdot 10^8$ mechanical quantum, the number of loads (wheel revolutions), equal to the critical number of loading cycles —63 millions, leads to fatigue wear (loss) of the material layer of unit thickness h_*. Linear wear h_* of the gear wheel is equal to the diameter $Q_{TS} = 2.85 \cdot 10^{-6}$ m of the equilibrium friction volume V_f^* (**Figure 10**) [19]. This is a physical criterion of wear. Accordingly, it is clear that the constructive limit criterion of wear of the tooth of the gear is equal to the limit of wear when the bending strength of the tooth is violated. For example, this is approximately 0.3 modulus of the tooth of the gear.

Consequently, the elementary nanostructure of deformable solids should be considered as the wear standard and used to optimize the operating time of real highly loaded Hertzian friction systems.

5.2 Evaluation of the capacity for work of bearings of internal combustion engines

Let's take an engine with an average shaft rotation—$n = 1500$ min^{-1}. Take the limit linear wear of the bearing which is equal to $h^* = 0.1$ mm. We know the linear size of the elementary tribosystem—$D_{TS} = 2.85$ mkm $= 2.85 \cdot 10^{-6} m$ [19]. For each revolution of the shaft one elementary tribosystem (equilibrium, run-in contact) loses one mechanical quantum. The number of revolutions required for the wear of one elementary tribosystem is equal to the number of mechanical quanta in this tribosystem, that is, it is $n_{MQ} = 0.63 \cdot 10^8$ revolutions.

Now we can determine the wear time of one elementary tribosystem:

$$t_{TS} = \frac{n_{MQ}}{n} = \frac{0.63 \cdot 10^8}{1500} = 42,000 \text{ min} = \frac{42,000}{60} = 700 \text{ hour} = \frac{700}{24} = 29,166 \text{ day} = \frac{29,166}{365} = 0.0799 \text{ year}. \quad (50)$$

Now let's define the number of layers of elementary tribosystems into linear wear—0.1 mm:

$$a_{h^*} = \frac{h^*}{D_{TS}} = \frac{1 \cdot 10^{-4}}{2.85 \cdot 10^{-6}} = 0.35 \cdot 10^2 = 35. \tag{51}$$

Now, let's define the time of wear of shaft-bearing system with the ultimate linear given wear—$h^* = 0.1$ mm, namely:

$$t_{motor} = t_{TS} \cdot a_{h^*} = 0.0799 \cdot 35 = 2,7968 \text{ year}. \tag{52}$$

Finally, we have 2, 7968 years of continuous work at ultimate load.

For this result we have the wear rate—$i = 4\, nm/h$. For example, this fits well with the data for the engine wear rate—$i = 5\, nm/h$ specified by Prof. F. Franek [20].

If we work 8 hours per day, then we will get the following result:

$$2,7968 \cdot 3 = 8.39 \text{ year}. \tag{53}$$

This is a real result for modern cars. If we work less than 8 hours a day, then the duration will increase significantly.

5.3 The principle of critical wheel rolling speed

The limit of this speed is determined by the principle of filling the entire nominal friction area of the sliding system with elementary tribosystems damping the process. Above this speed of movement of the vehicle there will be a complete unloading of the tribosystem, the separation of the wheel from the rail surface, since the principle of minimum resistance to movement (the principle of one elementary tribosystem or the principle of irreversibility) will be violated. In this case, all mechanical quanta in the elementary tribosystem will repel the wheel. There will be no quantum activating the process of maintaining the system in an excited state.

The calculation will be made in the following order [21]. The elementary nominal size of the contact area is known. By definition [4], $n_*^{TS} = 0.63 \cdot 10^8$ elementary tribosystems can be placed and operate on the elementary nominal contact area. Each elementary tribosystem (for the model of spherical roughness) has a size $D_{1TS} = 2.85 \cdot 10^{-6}\, m$ [19] and is capable of providing a rolling path of the wheel in the elementary act of rolling on the length of this tribosystem.

Thus, if all elementary tribosystems work in a unit of time on the entire nominal contact area, then the path traversed by the wheel in a unit of time is equal to

$$L_{\Sigma TS} = D_{1TS} \cdot n_{TS}^* = 2.85 \cdot 10^{-6} \cdot 0.63 \cdot 10^8 = 179.55\, m. \tag{54}$$

Consequently, the critical speed of wheel rolling is equal

$$\text{v}_* = L_{\Sigma TS} \cdot 3600 = 646,38\, km/h. \tag{55}$$

This result is close to modern speed of 574.8 km/h (TGV, France).

5.4 Self-organized nanoquantum solid lubricant

Information above allows us to consider new self-organized surface layer as follows: (1) the layer that separates the two original surfaces (alloys) of friction

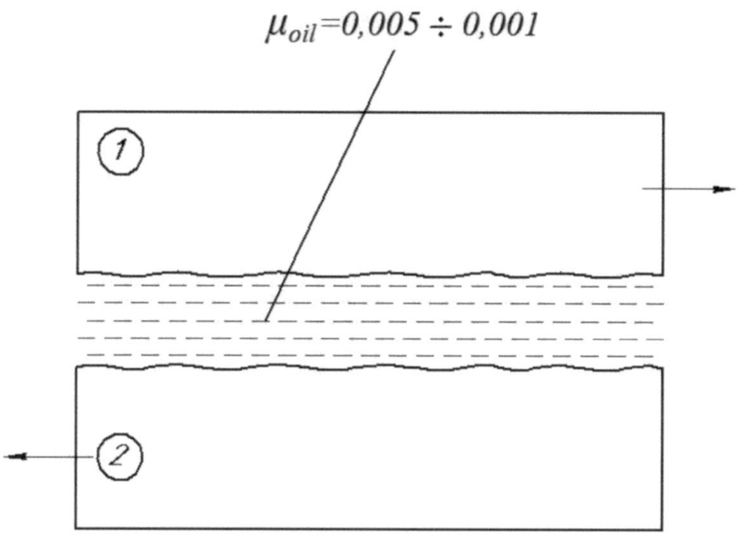

Figure 11.
Notional scheme of hydrodynamic lubrication.

from each other; (2) layer, which has a low coefficient of internal friction; (3) layer, which has a high capacity for work, that is, very small wear; and (4) layer, which may be seen as a solid lubricant.

Now you need to determine a value for the coefficient of friction of this self-organized solid lubricant and compare it with the coefficient of friction, for example, the most effective, or hydrodynamic lubrication.

It is known that the hydrodynamic lubrication when the stationary condition (**Figure 11**) has coefficients of friction μ down to 0.005÷0.001 values.

For nanoquantum self-organized solid lubricant friction coefficient will be calculated in the following order:

1. It is known [4, 15] that between the nanoquanta coefficient of friction is equal to $\mu_{MQ} = 1.587 \cdot 10^{-8}$.

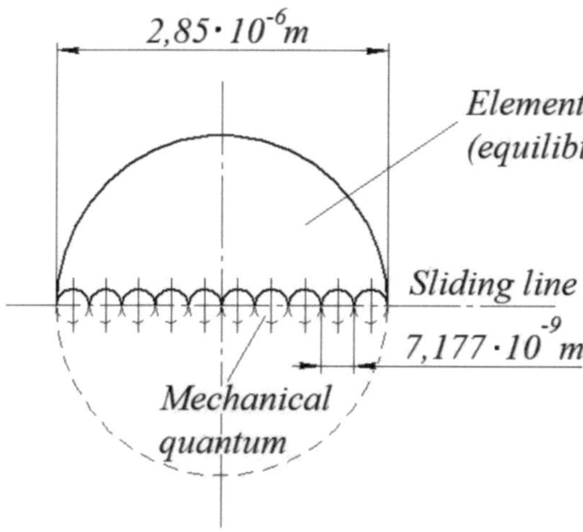

Figure 12.
Conditional scheme for equilibrium elementary tribosystem, structured mechanical quanta. At length of this tribosystem there are 397 mechanical quantums.

2. It is known [19] that the size of the critical volume of frictional contact (elementary tribosystem) is equal to $D_{TS} = 2.85 \cdot 10^{-6}\, m$.

3. Let's picture an elementary tribosystem in the plane as a circle with a diameter of $D_{TS} = 2.85\, mkm$ (**Figure 12**).

4. Next, let's define the number of mechanical (nano) quanta n'_{MQ} on a length D_{TS} of elementary tribosystem (**Figure 12**):

$$n'_{MQ} = \frac{D_{TS}}{D_{MQ}} = \frac{2.85 \cdot 10^{-6}}{7.177 \cdot 10^{-9}} = 397. \tag{56}$$

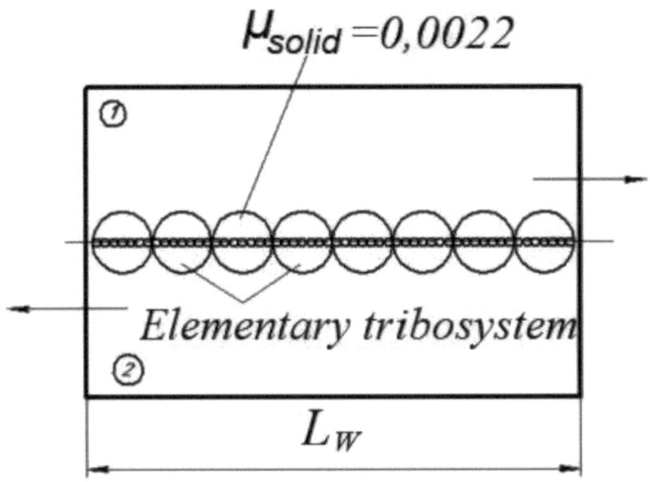

Figure 13.
Notional scheme of friction on the wavelength, structured elementary tribosystems. At the surface friction wavelength is 351 elementary tribosystems.

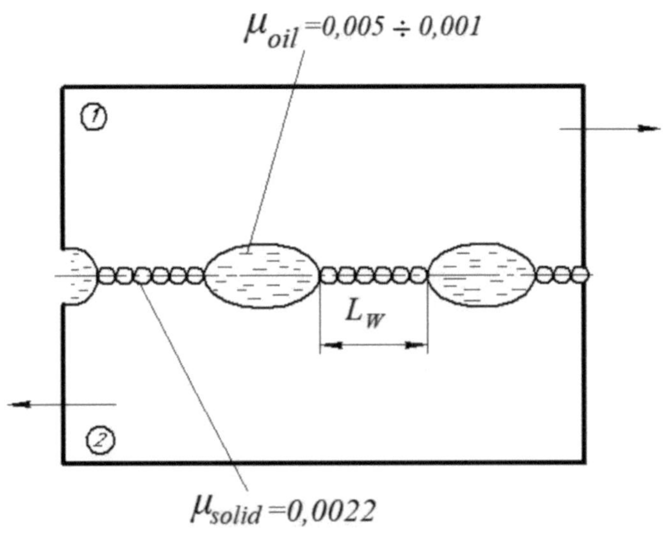

Figure 14.
Notional scheme of self-organized nanoquantum contact with unsteady hydrodynamic lubrication.

5. Let's define the coefficient of friction for a single equilibrium critical volume of friction (elementary tribosystem), the length of which is 397 mechanical quantums (**Figure 12**).

$$\mu_{TS} = \mu_{MQ} \cdot n'_{MQ} = 1.587 \cdot 10^{-8} \cdot 397 = 0.63 \cdot 10^{-5}. \tag{57}$$

6. Let's take the average friction surface wavelength equal to $L_W \cong 1 \cdot 10^{-3} m$.

Now define a number of elementary tribosystems on this wave length (**Figure 13**)

$$n_{TS} = \frac{L_W}{D_{TS}} = \frac{1 \cdot 10^{-3}}{2.85 \cdot 10^{-6}} = 351. \tag{58}$$

7. Now define friction coefficient at a wavelength of friction surface

$$\mu_W = \mu_{TS} \cdot n_{TS} = 0.63 \cdot 10^{-5} \cdot 351 = 0.0022 \tag{59}$$

As a result, we have a full conformity (**Figure 14**) of friction coefficient values for hydrodynamic lubrication—0.005÷0.001 and solid lubricant—0.0022.

Thus, it is fair to talk about nanoquantum self-organized solid lubrication.

6. Conclusions

1. Structural-energy analysis of the friction process allows us to consider the friction process as an evolutionary process.

2. From the equations of the energy balance of friction it follows that the evolution of the tribosystem (contact) has an adaptive-dissipative character.

3. The coefficient of friction has an energy interpretation that reveals its deep physical sense.

4. Experimental friction curves of $\mu = \mu(N, v)$ type may be examined as generalized friction experimental curves.

5. Structural-energy diagram of the evolution of rubbing surfaces (friction contact) interprets the general regularities of transformation and dissipation of energy during friction.

6. In the process of evolution of the friction contact, an elementary tribosystem is formed as a self-sufficient energy transformer under friction. This elementary tribosystem (critical friction volume) can be considered as an analogue of the material point of mechanics.

7. The most complete evolution of the tribosystem has a symmetrical form—the friction process begins and ends in the elastic region.

8. With the most complete evolution of the friction contact (elementary tribosystem), a unique nanostructure (tribosubsystem) is formed; the basis of this nanostructure is a mechanical (nano) quantum and the friction contact (material point of mechanics) consists of about $0.63 \cdot 10^8$ such nano quanta.

9. We can consider the mechanical quantum as the smallest structural form of a material solid and as the structural standard of material solid.

10. The mechanical quantum is precisely an asymptotically stable attractor of the limit cycle type for a deformable solid body (at friction).

11. All parameters of compatibility (optimal) friction have to be in quanta levels —commensurable with the parameters of the one mechanical quantum.

12. Interaction between nanoquantums is nature the net elasticity. The value of the coefficient of friction between mechanical quanta has order $\mu_{MQ} = 1.587 \cdot 10^{-8}$.

13. Exploitation of gear wheels and other heavy-loaded tribosystems (Hertzian contact) are subjected to model of nano-quantum damping, when one mechanical quantum is the standard of contact structure and wear.

Author details

Sergey Fedorov
Kaliningrad State Technical University, Kaliningrad, Russia

*Address all correspondence to: fedorov@klgtu.ru

References

[1] Fedorov VV. Thermodynamic Aspects of Strength and Fracture of Solids. Tashkent: Science; 1979. p. 168. (In Russian)

[2] Fedorov VV. Kinetics of Damage and Fracture of Solids. Tashkent: Science; 1985. p. 186. (In Russian)

[3] Fedorov VV. Ergodynamic concept of failure. In: Strength of Materials, Translated from Russian. Vol. 23/8. New York: Consultants Bureau; 1991. pp. 883-889

[4] Fedorov SV. The Foundations of Triboergodynamics and Physico-Chemical Prerequisits of Compatibility Theory. Kaliningrad: Kaliningrad State Technical University Press; 2003. p. 416. (In Russian)

[5] Ivanova VS. Synergetics: Strength and Destruction of Metallic Materials. Moscow: Science; 1992. p. 160. (In Russian)

[6] Fleischer G. Die Tross'schen Erkenntnisse aus heutigen Sicht. In: 1st Arnold Tross Kolloquium. Germany: Hamburg; 2005. pp. 215-242

[7] Fedorov SV. Physical and constructive (limiting) criterions of gear wheels wear. In: Proceedings of the 9th International Conference on Tribology (Balkantrib'17); IOP Publishing IOP Conf. Series: Materials Science and Engineering. Vol. 295. 2018. p. 012038. DOI: 10.1088/1757-899X/295/1/012038

[8] Fedorov SV. Nano-structural standard of friction and wear. Tribology in Industry. 2018;**40**(/2):225-238

[9] Fedorov SV. Structural-energetic regularities of tribocontact evolution. In: Proceedings of the 13th International Conference on Tribology, ROTRIB'16 IOP Publishing. IOP Conf. Series: Materials Science and Engineering. Vol. 174. 2017. p. 012012. DOI: 10.1088/1757-899X/174/1/012012

[10] Kragelskii IV, Dobychin MN, Kombalov VS. Friction and Wear Calculation Methods. Moscow: Mechanical Engineering; 1977. p. 526. (In Russian)

[11] Bowden FP, Tabor D. The Friction and Lubrication of Solids. Oxford: Oxford University Press; 2001. p. 424

[12] Fedorov SV. Energy balance of friction and friction coefficient in energetical interpretation. Tribology in Industry. 2015;**37**(3):380-389

[13] Conti P. Sulla Resistenza di Attrito. Vol. II. Royal Akademia dei Lencei; 1875

[14] Lancaster IK. Basic mechanisms of friction and wear of polymers. Plastics and Polymers. 1973;**41**:296-306

[15] Fedorov SV. Energy model of sliding friction coefficient and generalized regularities of tribosystems evolution. In: Proceeding of the International Colloquium Tribology-Industrial and Automotive Lubrication; 11-13 January 2014; TAE. Available from: www.tae.de info@tae.de

[16] Fedorov SV. The friction coefficient and its relation to the contact fatigue characteristics of materials. Industrial Laboratory. 1995;**61**(1):41-49

[17] Fedorov SV. Generalized energy model of sliding friction coefficient and regularities of tribosystem evolution. In: Proceedings of the V World Tribology Congress (V WTC 2013); 8-13 September 2013; Turino, Italy; 2013. ISBN 9788890818509

[18] Fedorov SV. The mechanical quantum of dissipative friction structures is the elementary

tribonanostructure. In: Proceedings of the IV World Tribology Congress (IV WTC 2009); 13-19 September 2009; Kyoto, Japan: 2009. p. 926

[19] Fedorov SV. Calculation of the true friction volume. Friction & Lubrication in Machines and Mechanisms. 2010;**5**: 3-7. (In Russian)

[20] Franek F, Wopelka T, Jech M. On-board applicable high-resolution wear measurement technique for internal combustion engines. In: Proceedings of the International Conference BALTTRIB'2011; 30-31 October; Kaunas. Luthuania: 2011. pp. 196-201

[21] Fedorov SV. Energetical nature of the wheel elastic rolling. In: Proceedings of the 7th International Conference (Contact Mechanics and Wear Of Rail/ Wheel Systems); 24-27 September 2006; Brisbane, Australia. Vol. 1. 2006. pp. 105-114

Chapter 4

Thin Films: Study of the Influence of the Micro-Abrasive Wear Modes on the Volume of Wear and Coefficient of Friction

Ronaldo Câmara Cozza

Abstract

The purpose of this work is to study the influence of the micro-abrasive wear modes on the behaviors of the volume of wear (V) and of the coefficient of friction (μ) of thin films submitted to micro-abrasive wear. Experiments were conducted with thin films of TiN, TiAlN, TiN/TiAlN, TiHfC, ZrN, and TiZrN, using a ball of AISI 52100 steel and abrasive slurries prepared with black silicon carbide (SiC) particles and glycerine. The results show that the abrasive slurry concentration affected the micro-abrasive wear modes ("grooving abrasion" or "rolling abrasion") and, consequently, the magnitude of the volume of wear and of the coefficient of friction, as described: (i) a low value of abrasive slurry concentration generated "grooving abrasion," which was related to a relatively low volume of wear and high coefficient of friction, and (ii) a high value of abrasive slurry concentration generated "rolling abrasion," which was related to a relatively high volume of wear and low coefficient of friction.

Keywords: micro-abrasive wear, grooving abrasion, rolling abrasion, thin films, volume of wear, coefficient of friction

1. Introduction

The micro-abrasive wear test by rotating ball ("ball-cratering wear test") is an important method adopted to study the micro-abrasive wear behavior of metallic, polymeric, and ceramic materials. **Figure 1** presents a schematic diagram of the principle of this micro-abrasive wear test, in which a rotating ball is forced against the tested specimen in the presence of an abrasive slurry, generating, consequently, the called "wear craters" on the surface of the tested material.

Initially, the development of the ball-cratering wear test aimed to measure the thickness of thin films (**Figure 2a** and **b**) [1], which can be made using the equations detailed in Ref. [2]. Because of the technical features, this type of micro-abrasive wear test has been applied to study the tribological behavior of different materials [3–5], for example, in the analysis of the volume of wear (V), coefficient of wear (k), and coefficient of friction (μ) of thin films [2, 6–10].

As a function of the abrasive slurry concentration, two micro-abrasive wear modes can be usually observed on the surface of the worn crater: "grooving abrasion" is observed when the abrasive particles slide on the surface, whereas "rolling abrasion" results from abrasive particles rolling on the specimen's surface.

Figure 3a [11, 12] and **Figure 3b** presents, respectively, images of "grooving abrasion" and "rolling abrasion."

Many works on coefficient of friction (μ) during abrasive wear and other types of tests are available in the literature [13–19], but only a few were dedicated to the coefficient of friction in ball-cratering wear tests [2–4, 10, 11]. In particular, Shipway [20] has studied the coefficient of friction in terms of the shape and movement of the abrasive particles, Kusano and Hutchings [21] presented a theoretical model for coefficient of friction in micro-abrasive wear tests with "free-ball" equipment configuration, and Cozza et al. [2–4, 11, 22] measured the tangential force developed during tests conducted in a "fixed ball" equipment configuration, which allowed direct calculation of the friction coefficient by the ratio between the tangential and normal forces. Besides, using a proper electronic instrumentation, Cozza et al. [2, 23–25] have studied and measured the behavior of the coefficient of friction in thin films in ball-cratering wear tests; however, in those works [2, 23–25], the test sphere has reached the substrate.

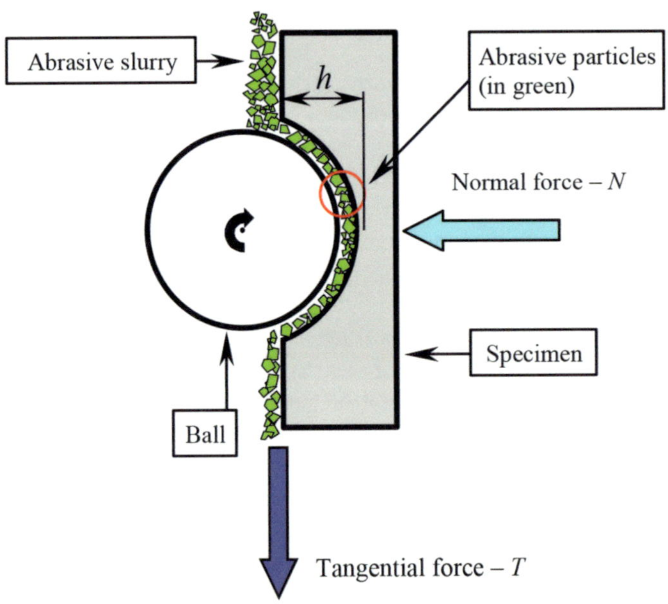

Figure 1.
Micro-abrasive wear test by rotating ball: a representative figure showing the operating principle and the abrasive particles between the ball and the specimen; "h" is the depth of the wear crater.

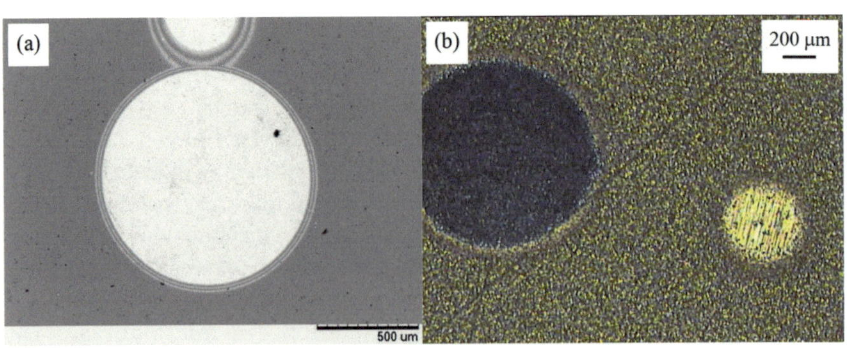

Figure 2.
Examples of wear craters generated on coated system: (a) multilayer and (b) thin film of TiN.

Figure 3.
Micro-abrasive wear modes: (a) "grooving abrasion" [11, 12] and (b) "rolling abrasion."

Analyzing and studying important researches regarding to tribological behavior of materials submitted to micro-abrasive wear test conditions [7–9, 26], the purpose of this work is to report the influence of the micro-abrasive wear modes on the behaviors of the volume of wear (V) and coefficient of friction (μ) of thin films submitted to micro-abrasive wear tests by rotating ball.

2. Equipment, materials, and methods

2.1 Ball-cratering wear test equipment

A ball-cratering wear test equipment with free-ball mechanical configuration (**Figure 4** [27]) was used for the micro-abrasive wear tests, which has two load cells: one load cell to control the "normal force" (N) and one load cell to measure the "tangential force" (T) that is developed during the experiments. The values of "N" and "T" are read by a readout system.

2.2 Materials

Experiments were conducted with thin films of:

- TiN
- TiAlN
- TiN/TiAlN
- TiHfC
- ZrN
- TiZrN

deposited on substrates of cemented carbide. For the counter-body, one ball of AISI 52100 steel with diameter of D = 25.4 mm was used.

The abrasive material was black silicon carbide (SiC) with an average particle size of 3 μm; **Figure 5** [4] presents a micrograph of the abrasive particles (**Figure 5a**)

Figure 4.
Ball-cratering micro-abrasive wear test equipment used in this work: free-ball mechanical configuration, able to acquire, simultaneously, the "normal force N*" and the "tangential force* T*."*

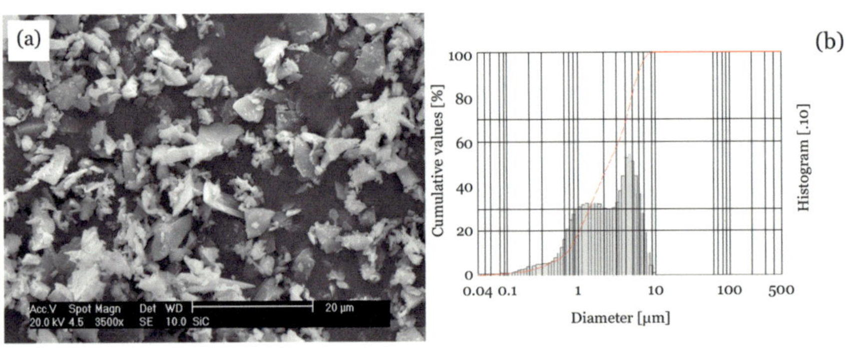

Figure 5.
SiC abrasive [4]: (a) scanning electron micrograph and (b) particle size distribution.

Normal force [N]	N	0.4
Abrasive slurry concentration (in volume)	C_1	5% SiC + 95% glycerine
	C_2	50% SiC + 50% glycerine
Ball rotational speed [rpm]	n	70

Applied on thin films of:

TiN

TiAlN

TiN/TiAlN

TiHfC

ZrN

TiZrN

Table 1.
Test parameters selected for the ball-cratering wear experiments.

and the particle size distribution (**Figure 5b**). The abrasive slurries were prepared with SiC and glycerine.

2.3 Methods

Table 1 presents the values of the test parameters defined for the micro-abrasive wear experiments.

The normal force value defined for the wear experiments was N = 0.4 N, combined with two abrasive slurries concentrations (C), C_1 = 5% SiC + 95% glycerine and C_2 = 50% SiC + 50% glycerine (volumetric values), with the purpose to produce, respectively, "grooving abrasion" and "rolling abrasion" on the surfaces of the thin films. The ball rotational speed was set to n = 70 rpm.

All tests were *non-perforating*, e.g., only the thin films were worn. The normal force (N) was constant during the tests; the tangential force (T) was monitored and registered during all experiments.

The volume of wear (V) and the coefficient of friction (μ) were then calculated using Eqs. (1) [1] and (2), respectively; "d" is the diameter of the wear crater, and "R" is the radius of the ball:

$$V \approx \frac{\pi d^4}{64R}, \text{for } d < < R \quad (1)$$

$$\mu = \frac{T}{N} \quad (2)$$

3. Results and discussion

Figures 6 and 7 show examples of worn surfaces obtained in the experiments; in all wear craters, the maximum depth (h) observed was, approximately, $h \approx 8$ μm. **Figure 6** displays the action of "grooving abrasion," characteristic of C_1 = 5%

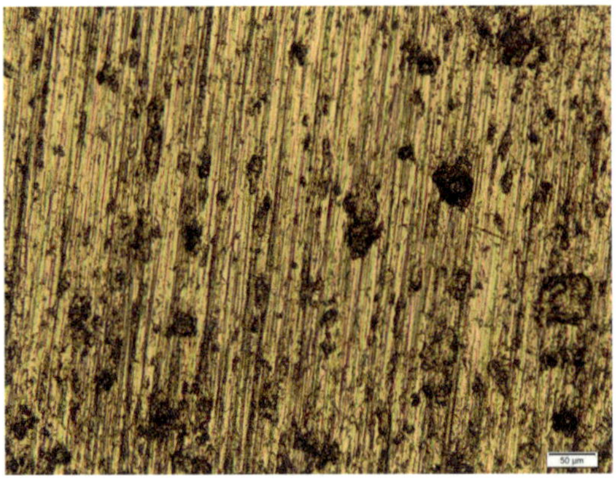

Figure 6.
Occurrence of "grooving abrasion" on the surface of the thin film of TiN.

Figure 7.
Occurrence of "rolling abrasion" on the surface of the thin film of TiN.

SiC + 95% glycerine; **Figure 7** displays a wear crater under the action of "rolling abrasion," reported for the abrasive slurry concentration C_2 = 50% SiC + 50% glycerine. These results qualitatively agree with the conclusions obtained by Trezona et al. [28], in which low concentrations of abrasive slurries (<5% in volume of abrasive material, approximately) favor the occurrence of "grooving abrasion" and high concentrations of abrasive slurries (>20% in volume of abrasive material, approximately) favor the action of "rolling abrasion."

The actions of the micro-abrasive wear modes showed an important influence on the volume of wear and on the coefficient of friction of the thin films studied in this research. A significant increase in the volume of abrasive particles from C_1 = 5% SiC + 95% glycerine to C_2 = 50% SiC + 50% glycerine (causing, consequently, the micro-abrasive wear transition from "grooving abrasion" to "rolling abrasion") caused an increase in the volume of wear and a decrease in the coefficient of friction.

Figures 8 and **9** show the behaviors of the volume of wear (V) and coefficient of friction (μ) as a function of the micro-abrasive wear modes; the maximum errors observed were $V = 0.4 \times 10^{-3}$ mm^3 and $\mu = 0.1$, for the volume of wear and coefficient of friction, respectively.

The values of the volume of wear reported under conditions of "rolling abrasion" (high-abrasive slurry concentration, C_2 = 50% SiC + 50% glycerine) were higher than the values of the volume of wear reported under conditions of "grooving

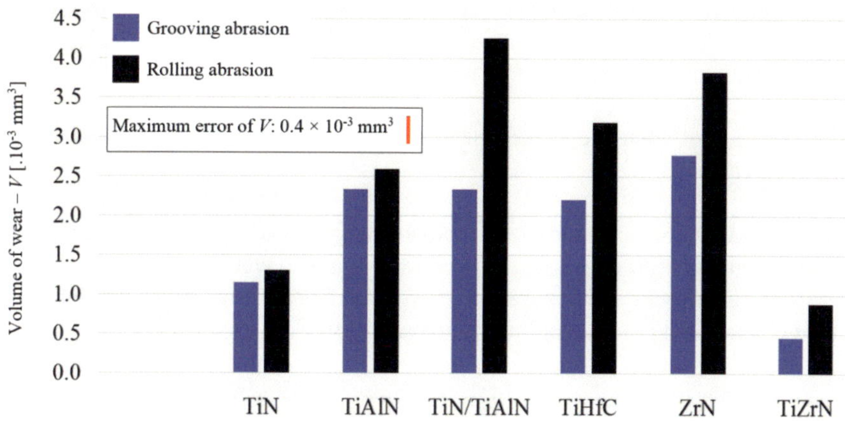

Figure 8.
Volume of wear (V) as a function of the micro-abrasive wear modes' "grooving abrasion" and "rolling abrasion."

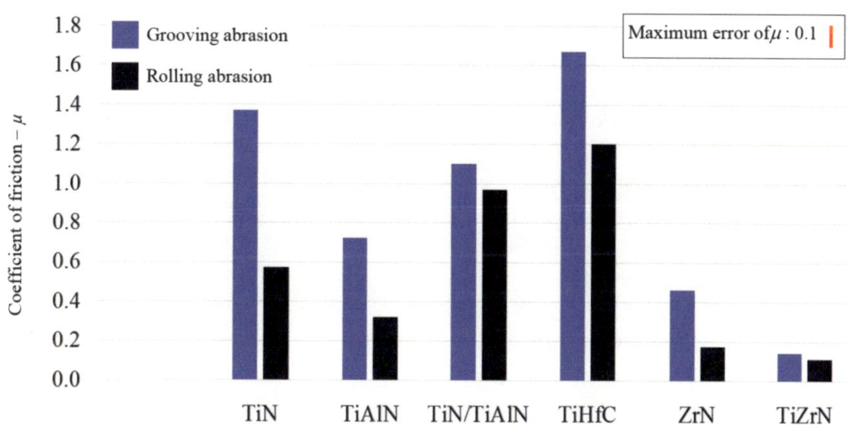

Figure 9.
Coefficient of friction (μ) as a function of the micro-abrasive wear modes' "grooving abrasion" and "rolling abrasion."

abrasion" (low-abrasive slurry concentration, C_1 = 5% SiC + 95% glycerine), as reported by Mergler and Huis in 't Veld [5] and Trezona et al. [28].

The values of the coefficient of friction reported under "grooving abrasion" (low-abrasive slurry concentration, C_1 = 5% SiC + 95% glycerine) were higher than the values of the coefficient of friction reported under "rolling abrasion" (high-abrasive slurry concentration, C_2 = 50% SiC + 50% glycerine), and this behavior can be explained based on patterns of movements that act on "rolling abrasion" and "grooving abrasion" micro-abrasive wear modes: in "rolling abrasion," the abrasive particles are free to roll between the ball and the specimen, facilitating the relative movement between these elements and, consequently, decreasing the coefficient of friction on the tribological system; however, in "grooving abrasion," the abrasive particles are fixed on the counter-body (in this case, on the ball), limiting their movements and requiring higher tangential forces.

4. Conclusions

The results obtained indicated the conclusions:

1. The concentration of abrasive slurry affected the occurrence of “grooving abrasion”—under low concentration—or “rolling abrasion,” under high concentration.

2. The volume of wear increased with the increase of the abrasive slurry concentration.

3. With the low concentration of abrasive slurry, “grooving abrasion” and, consequently, high values of coefficient of friction were reported. In this situation, the abrasive particles were incrusted on the counter-body, hindering their movements and generating high tangential forces.

4. On the other hand, when the high concentration of abrasive slurry was used, “rolling abrasion” occurred. In this case, the abrasive particles were free to roll along the surface of the thin film, causing a low coefficient of friction.

Appendix

A list of symbols used in this manuscript is given:

C	abrasive slurry concentration—in volume, [% SiC + % glycerine]
d	diameter of the wear crater, [mm]
D	diameter of the ball, [mm]
h	depth of the wear crater, [μm]
k	coefficient of wear, [mm^3/N m]
n	ball rotational speed, [rpm]
N	normal force, [N]
R	radius of the ball, [mm]
T	tangential force, [N]
V	volume of wear, [mm^3]
Greek letter	
μ	coefficient of friction

Author details

Ronaldo Câmara Cozza[1,2]

1 Department of Mechanical Engineering, University Center FEI—Educational Foundation of Ignatius "Padre Sabóia de Medeiros", São Bernardo do Campo, SP, Brazil

2 Department of Mechanical Manufacturing, CEETEPS—State Center of Technological Education "Paula Souza", Technology Faculty—FATEC, Mauá, SP, Brazil

*Address all correspondence to: rcamara@fei.edu.br

References

[1] Rutherford KL, Hutchings IM. Theory and application of a micro-scale abrasive wear test. Journal of Testing and Evaluation. 1997;**25**(2):250-260

[2] Cozza RC. A study on friction coefficient and wear coefficient of coated systems submitted to micro-scale abrasion tests. Surface and Coatings Technology. 2013;**215**:224-233

[3] Cozza RC, Tanaka DK, Souza RM. Friction coefficient and abrasive wear modes in ball-cratering tests conducted at constant normal force and constant pressure—Preliminary results. Wear. 2009;**267**:61-70

[4] Cozza RC, Rodrigues LC, Schön CG. Analysis of the micro-abrasive wear behavior of an iron aluminide alloy under ambient and high-temperature conditions. Wear. 2015;**330-331**:250-260

[5] Mergler YJ, Huis in 't Veld AJ. Micro-abrasive wear of semi-crystalline polymers. Tribology and Interface Engineering Series. 2003;**41**:165-173

[6] Cozza RC, Tanaka DK, Souza RM. Micro-abrasive wear of DC and pulsed DC titanium nitride thin films with different levels of film residual stresses. Surface and Coatings Technology. 2006;**201**:4242-4246

[7] Batista JCA, Godoy C, Matthews A. Micro-scale abrasive wear testing of duplex and non-duplex (single-layered) PVD (Ti, Al)N, TiN and Cr-N coatings. Tribology International. 2002;**35**:363-372

[8] Schiffmann KI, Bethke R, Kristen N. Analysis of perforating and non-perforating micro-scale abrasion tests on coated substrates. Surface and Coatings Technology. 2005;**200**:2348-2357

[9] Kusano Y, van Acker K, Hutchings IM. Methods of data analysis for the micro-scale abrasion test on coated substrates. Surface and Coatings Technology. 2004;**183**:312-327

[10] Rutherford KL, Hutchings IM. A micro-abrasive wear test, with particular application to coated systems. Surface and Coatings Technology. 1996;**79**:231-239

[11] Cozza RC. Influence of the normal force, abrasive slurry concentration and abrasive wear modes on the coefficient of friction in ball-cratering wear tests. Tribology International. 2014;**70**:52-62

[12] Cozza RC. Third abrasive wear mode: Is it possible? Journal of Materials Research and Technology. 2014;**3**(2):191-193

[13] Neville A, Kollia-Rafailidi V. A comparison of boundary wear film formation on steel and a thermal sprayed Co/Cr/Mo coating under sliding conditions. Wear. 2002;**252**:227-239

[14] Dhanasekaran S, Gnanamoorthy R. Abrasive wear behavior of sintered steels prepared with MoS_2 addition. Wear. 2007;**262**:617-623

[15] Ceschini L, Palombarini G, Sambogna G, Firrao D, Scavino G, Ubertalli G. Friction and wear behaviour of sintered steels submitted to sliding and abrasion tests. Tribology International. 2006;**39**:748-755

[16] Gheisari R, Polycarpou AA. Three-body abrasive wear of hard coatings: Effects of hardness and roughness. Thin Solid Films. 2018;**666**:66-75

[17] Atapour M, Blawert C, Zheludkevich ML. The wear characteristics of CeO_2 containing nanocomposite coating made by aluminate-based PEO on AM 50 magnesium alloy. Surface & Coatings Technology. 2019;**357**:626-637

[18] Geng Z, Shi G, Shao T, Liu Y, Duan D, Reddyhoff T. Tribological behavior of patterned TiAlN coatings at elevated temperatures. Surface & Coatings Technology. 2019;**364**:99-114

[19] Ghasemi MH, Ghasemi B, Semnani HRM. Wear performance of DLC coating on plasma nitrided Astaloy Mo. Diamond and Related Materials. 2019;**93**:8-15

[20] Shipway PH. A mechanical model for particle motion in the micro-scale abrasion wear test. Wear. 2004;**257**:984-991

[21] Kusano Y, Hutchings IM. Sources of variability in the free-ball micro-scale abrasion test. Wear. 2005;**258**:313-317

[22] Cozza RC. Wear and coefficient of friction study in micro-abrasive wear tests with rotating ball under conditions of "constant normal force" and "constant pressure" [Ph.D. thesis]. São Paulo, SP, Brazil: Polytechnic School of the University of São Paulo; 2011. p. 327. Available from: http://www.teses.usp.br/

[23] Cozza RC, Wilcken JTSL, Delijaicov S, Donato GHB. Tribological characterization of thin films based on residual stress, volume of wear, micro-abrasive wear modes and coefficient of friction. In: Proceedings of the "ICMCTF 2017—44th International Conference on Metallurgical Coatings and Thin Films"; 24-28 April 2017; San Diego, California, USA. 2017

[24] Cozza RC, Donato GHB. Study of the influence of the abrasive slurry concentration on the coefficient of friction of thin films submitted to micro-abrasive wear. In: Proceedings of the "PacSurf 2016—Pacific Rim Symposium on Surfaces, Coatings & Interfaces"; 11-15 December 2016; Kohala Coast, Hawaii, USA. 2016

[25] Cozza RC, Wilcken JTSL, Schon CG. Influence of abrasive wear modes on the volume of wear and coefficient of friction of thin films. In: Proceedings of the "CoSI 2015—11th Coatings Science International"; 22-26 June 2015; Noordwijk, The Netherlands. 2015

[26] Batista JCA, Joseph MC, Godoy C, Matthews A. Micro-abrasion wear testing of PVD TiN coatings on untreated and plasma nitride AISI H13 steel. Wear. 2002;**249**:971-979

[27] Cozza RC. Effect of sliding distance on abrasive wear modes transition. Journal of Materials Research and Technology. 2015;**4**(2):144-150

[28] Trezona RI, Allsopp DN, Hutchings IM. Transitions between two-body and three-body abrasive wear: Influence of test conditions in the microscale abrasive wear test. Wear. 1999;**225-229**:205-214

Chapter 5

Novel Predictors for Friction and Wear in Drivetrain Applications

Walter Holweger

Abstract

Reliability in a drivetrain is given by the life of its constituents, e.g., gears, clutches, and bearings. Lubrication contributes to the life cycle, preventing wear, friction, and environmental impacts. As lubricants and their additives are chemicals with an expected reactivity in a tribological contact, it comes to the question how surface fatigue phenomena due to loading may be influenced by the reactivity of functional additives and how this might be embedded in construction guidelines. A very basic study based on an elementary gear test rig presents the result that pitting life of a gear is substantially influenced by the chemical structure of wear-preventing additives. Even under appropriate loading conditions, the lubricant structure comes as a life-limiting factor. A molecular model shows how the release and the approach of the additives toward a surface is essential and related to the reaction processes that occur during the loading.

Keywords: drivetrain, gears, bearings, reliability, pitting, wear, gray staining, life cycle, molecular modeling

1. Introduction

Wear is a central topic in tribology. As a system property, it is defined as a continuous loss of material out of a solid surface, caused by mechanical impact, e.g., contact and relative motion of counterpart such as solids, liquids, or gases [1–6].

As such, wear is not a property of a single component. Drivetrain components (e.g., bearings, gears, clutches, etc.) are constructed due to their life expectation in order to come to a predictive reliability in the life cycle. However, in reality they are exposed to wear processes as an incidental or continuous impact. Hence, it is important to know how the entrance of wear in drivetrain components will influence their life expectations and the reliability of the drivetrain as such.

Within a construction, the expected life is a function of the load capacity of the materials, e.g., their fatigue strength with respect to load cycles and pressure.

As reliability is defined yet by the load capacity of the involved materials due to cyclic stress, the question is about how wear relates to fatigue. In a classical view, fatigue is a matter related to stress-strain properties due to the elastic plastic behavior of the load carrying components. If a pressure with no tangential component acts on moving parts, the fatigue phenomena are given by slow changes of the subsurface microstructure due to phase alterations, migration of interstitial atoms, and dislocations. As tangential forces due to slip are coming up, the fatigue processes moves up toward the surface. However, fatigue phenomena near the surface will bring up the question at which point fatigue crosses wear and vice versa. While reliability up to now is defined by fatigue properties of the material, the crossing between fatigue and wear, especially those, induced by lubricants is still not solved. Within real applications it might be the case that, due to the operating conditions, fatigue comes to lubricant-induced wear and does not fit with the standard construction guidelines.

We present here a basic study, how fatigue and lubricant-induced wear push each other in a standard gear and bearing test. It comes up that this stimulation is due to the basic behavior of lubricant components, e.g., the reactivity of additives combined with the mechanical loading. As a main and future question of research, it addresses the need of advanced understanding on a molecular scale (10^{-9} m), molecular modeling, and in situ spectrometry to embed them in future construction guidelines.

2. Gear and bearing life in terms of lubricants

Pitting and gray staining in gears and bearings appear as surface features. In a worst case, they may promote a decay in life expectation, due to their progression in time.

Within the traditional view, they are interpreted by the assumption that loading exceeds the load capacity of the material. Consequently the mating parts will get in touch and come to rupture. As such, lubricants as separating media are only seen as a material to avoid this by separating the surfaces due to viscous effects. However, it is well known that lubricants as a matter of their composition will influence the surface load capacity as well (see **Figure 1**) [7, 8] as seen for gears in FZG standard test conditions, using SAE 4320 case-hardened material [7–9].

Figure 2 shows the wear rate by the use of different anti-wear and extreme pressure additives base on the FZG test rig (16) as a function of the pitch line speed:

Same as for gears, bearings are impacted also by wear raising from the composition of a lubricant [10, 11] (see **Figure 3**), using the Schaeffler FE8 test rig as a standard (2100 MPa contact pressure, 80 rpm, 80 h, cylindrical roller bearing, SAE 52100, Martensite):

Within the FZG gear test rig [9, 12–17] (DIN ISO 14635), different lubricants (A, B) differ in wear as a fact of temperature. While oil A shows a decay by raising the temperature, oil B is opposite (see **Figure 4**).

As a result of those studies, reaction layers with different thicknesses under mechanical influence are created. While thick and uncontrolled layers cause early fatigue and wear, thin oxide layers with a strong bonding to the interface cause no wear, same as reported earlier [1, 11]. It is of interest to describe these effects with respect to their chemical structures of the reactive components and how they undergo a transformation of the tribological contact area by creating those layers. Structure property relationship would lead to predictors for wear derived from the chemical structure of a given lubricant.

As a standard the FZG test rig (DIN ISO 14635) as a back-to-back gear test is used (**Figure 5**) [7, 18–20]. The gears, type FZG C-PT, are set in a gearbox, fully lubricated. Cylindrical roller bearings (type NJ406, steel cage) are used for the pinion shaft 1 and cylindrical roller bearings, type NJ308, for the motor shaft. Investigations were made on the gears and the cylindrical roller bearings NJ 406.

The test conditions are given in **Table 1**. The oil temperature is set constant to 90°C and motor speed to 1500 rpm. A running-in period with 1025 N/mm^2 is set for 2 h; the test run at 1700 N/mm^2 till pitting is reached is recorded. The speed at the

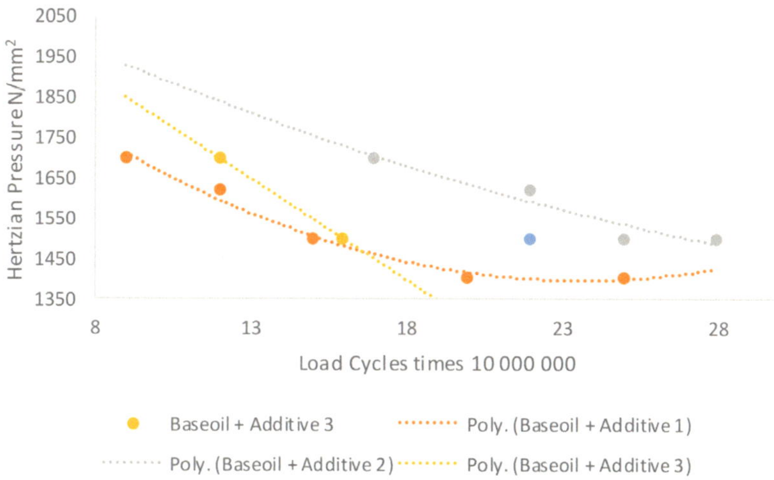

Figure 1.
The influence of different oils and additives on gear load cycles referring to the FZG test (DIN ISO 14635) [9].

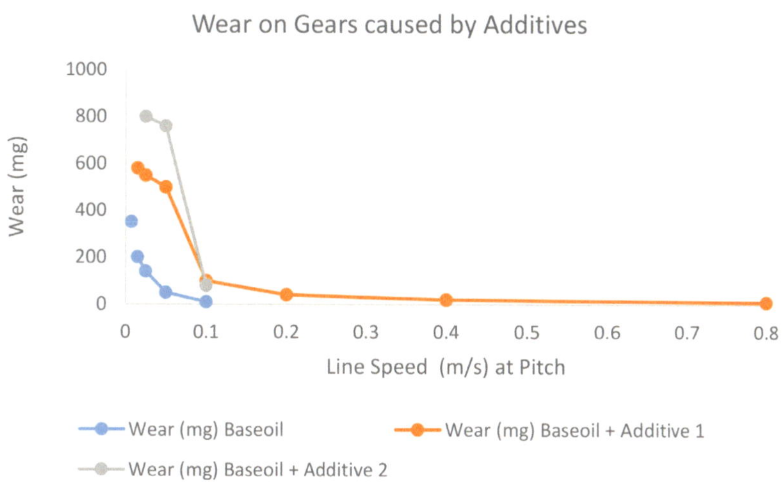

Figure 2.
Wear rate of lubricants as a function of pitch line speed.

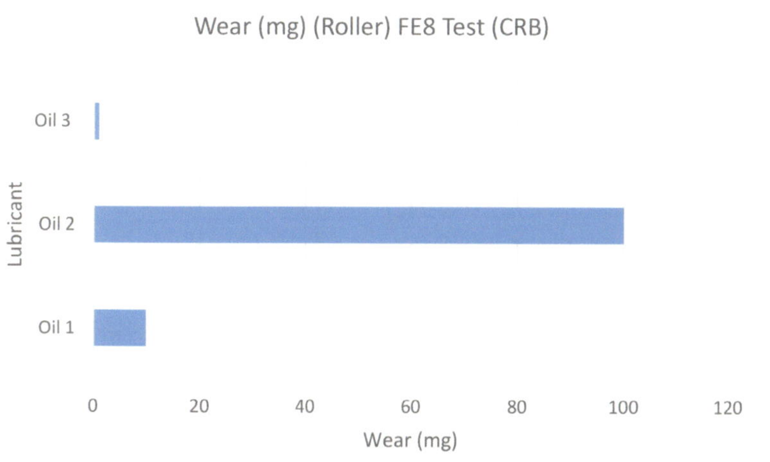

Figure 3.
Wear rate (roller) at a cylindrical roller bearing (CRB) from the Schaeffler test rig FE8 (DIN 51819) as a function of lubricants. While oil 1 and oil 3 do not show any wear, oil 2 is high in wear.

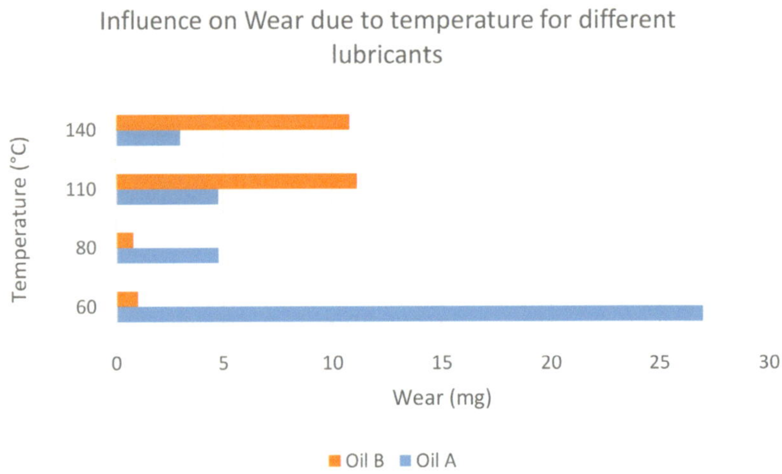

Figure 4.
Influence on wear due to temperature.

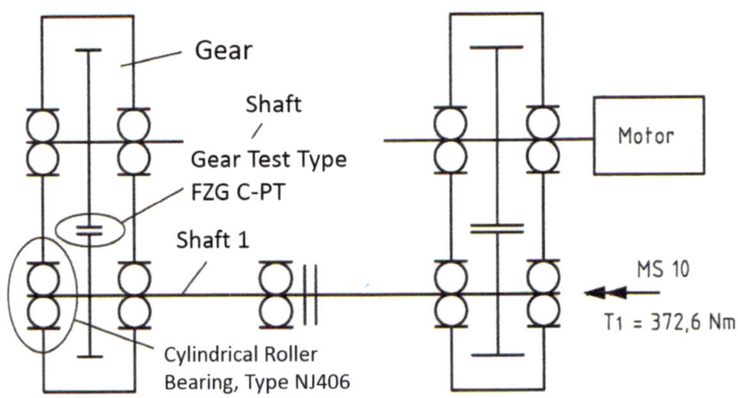

Figure 5.
FZG test rig (DIN ISO 14635).

Oiltemperature	90°C
Motor Speed	1500rpm
Running -in	1025 N/mm^2 for 2 hours
Test Run	1700 N/mm2
End of Test	Pitting, maximum 300 hours
Speed at the pinion	2250 rpm
Torque moment at the pinion	372,6 Nm
Type of Lubrication	Sump
Tagential Speed at the pinion	2.42 m/s
Tangential Speed at the wheel	3.87 m/s
Sliding speed at the pinion	-1.45 m/s
Sliding speed at the wheel	1.45 m/s
Sum of Speed	6.29 m/s

Table 1.
Conditions of the test.

pinion is set to 2250 rpm, the torque moment T1 to 372.6 Nm. The tangential speed at the pinion is calculated to 2.42 m/s, at the wheel to 3.87 m/s, the sliding speed at the pinion to −1.45 m/s (reflecting the negative slip), the sliding speed at the wheel to 1.45 m/s, and the sum of speed to 6.29 m/s. As the slip percentage is given by the ratio of sliding speed to the sum of the speed, the slip at the pinion is −23% and at the wheel +23%.

The material of the gears applies for a case-hardener SAE 4320.

The test specific data of the CRB NJ206 are given in **Table 2**.

The material of the bearing accords to the *SAE 52100*, martensitic hardening, tempered at 180°C, 2 hours, with 10–12% retained austenite.

Two lubricants were tested (**Table 3**). *Lubricant 1* reflects a standard technology, using zincdithiophosphates (C4ZndtP) as a sulfur-phosphorus carrier. As a representative of a new ashless additive technology, the *lubricant formulation 2* (C4NdtP) is used. The Poly-α-olefine viscosity is 46 mm^2/s at 40°C.

The organic chain length of the phosphorus-sulfur core is given by four C atoms, meaning that during the synthesis of the additives, a C4 (butyl) alcohol component was used.

The structure of the additives are shown in **Figures 6** and 7, both looking rather complex. In detail a core of sulfur, phosphorus, and zinc is attached to the carbon sites, containing four C atoms (ZndtPC4) (**Figure 6**).

Figure 7 represents the C4NdtP; two structures are held together by an ionic bonding: a sulfur-phosphorus component with two carbon sites, each containing four C atoms and their attached hydrogen and nitrogen component with a positively charged nitrogen at the edge, attached to a carbon site with eight C atoms (C4NdtP). The principal of this substance is similar to ionic liquids, where opposite-charged atoms create an ionic binding, while the carbon sites are responsible for the liquid structure.

The test runs by the use of the different additives are given in **Table 4** for both gears and bearings (NJ406) as a function of the load cycles. Clearly the table shows how the change in the chemical structure of the additive, despite the same chain lengths on the carbon edge (C4), end up in different load cycles (**Table 4**):

Position	Value
Innerdiameter	30 mm
Outerdiameter	90 mm
Width	23 mm
Diameter of the rollers	14 mm
Length of the rollers	11.6 mm
Curvature at Inner Race	22.5 mm
Curvature at Outer Race	-36.5 mm
Number of Rollers	9

Table 2.
Data from the CRB NJ206 bearing.

1	Poly-α-Olefine, ISO VG 46	99.29	%
	Zincdithiophosphate C4	0.71	%
2	Poly-α-Olefine, ISO VG 46	99.17	
	Ammoniumdithiophosphate C4	0.83	%

Table 3.
Lubricants used for the test.

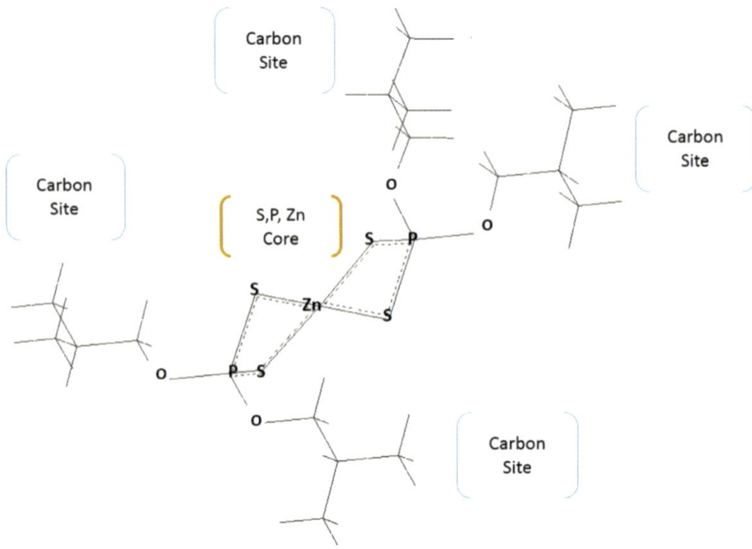

Figure 6.
Zincalkyldithiophosphate (C4ZndtP).

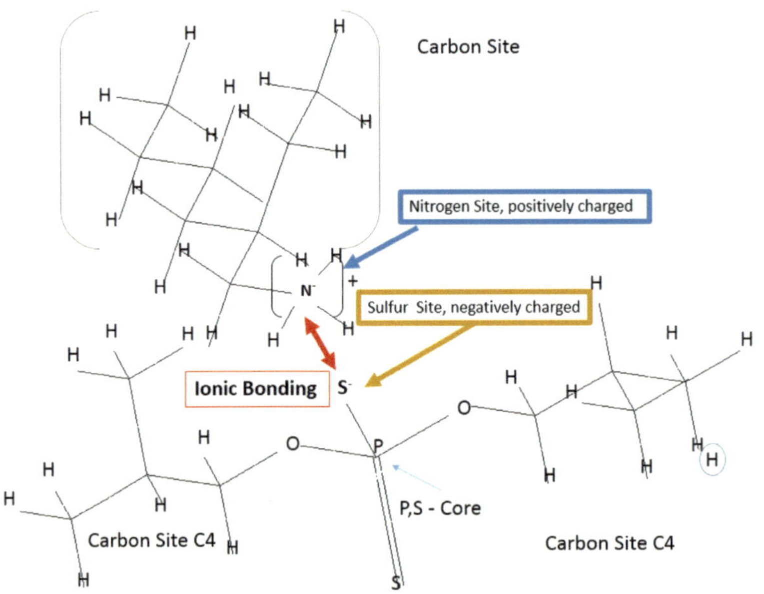

Figure 7.
Ammoniumdithiophosphate (C4NdtP) as an ionic liquid-like structure.

Run Nr	Component	Baseoil	Additive	Amount (%)	Temperature (°C)	Load Cycles(10^6)
1	Gear	Poly-α-Olefine, ISO VG 46	ZndtPC4	0.71	90	9.31
2	Gear	Poly-α-Olefine, ISO VG 46	ZndtPC4	0.71	90	7.8
3	Bearing	Poly-α-Olefine, ISO VG 46	ZndtPC4	0.71	90	
4	Gear	Poly-α-Olefine, ISO VG 46	NdtPC4	0.83	90	16.29
5	Gear	Poly-α-Olefine, ISO VG 46	NdtPC4	0.83	90	12.71
6	Bearing	Poly-α-Olefine, ISO VG 46	NdtPC4	0.83	90	

Table 4.
Test conditions set on the different additive structures.

3. Results for the gear

While the C4-Zincalkyldithiophosphate (C4ZndtP) causes pitting and does not meet the expected load cycles, the test carried out with the C4NdtP was out of failure [7]. Secondary neutral mass spectrometry (SNMS) profiles [21–23] were carried out at the pinion *addendum* (*Position 1*: see arrow in **Figure 8**) (area of *positive* slip referring to the pinion), the *pitch line* (*Position 2*: see arrow in **Figure 8**) (*zero slip* referring to the pinion), and tooth *dedendum* (*Position 3*: see arrow in **Figure 8**) (area of *negative slip*) in order to evaluate how the reaction rate of additives might depend on load cycles. The nature of the reaction was analyzed by secondary neutral mass spectrometry (SNMS). While secondary ion mass spectrometry (SIMS) is sensitive due to the local elements, specifically oxygen, SNMS is less sensitive and allows to track elements quantitatively as depth profiles from the top of the surface down to a few microns. The spatial resolution is around 4 mm^2, thus averaging local deviations in elements making the results more accurate.

The relevant depth profiles were taken at the dedendum of the pinion tooth flank for the additives C4-zincalkyldithiophosphate (C4ZndtP) and C4-aminealkyldithiophosphate (C4NdtP) with respect to load cycles are shown in **Figure 9** (C4ZndtP: 9×10^6 load cycles); **Figure 10** (C4ZndtP: 10×10^{10} load cycles); **Figure 11** (C4NdtP: 12×10^6 load cycles); and **Figure 12** (C4ZndtP: 16×10^{10} load cycles).

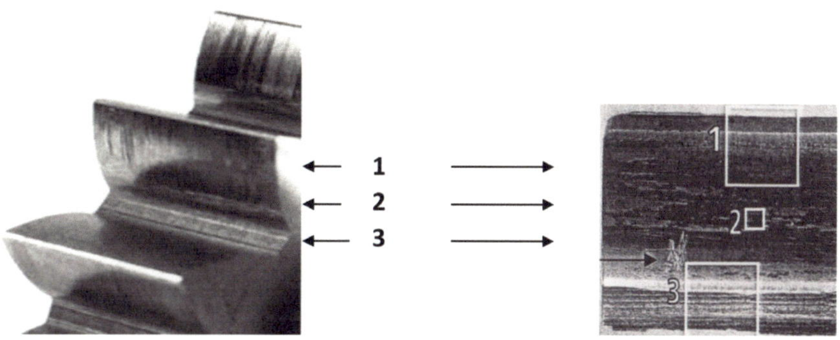

Figure 8.
Gear tooth segment with addendum Position 1 (pitch line), Position 2, and Position 3 as dedendum.

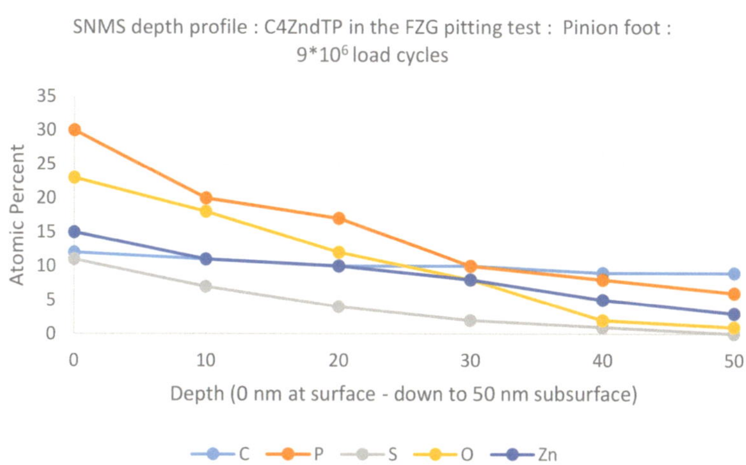

Figure 9.
SNMS depth profile: C4ZndtP in FZG pitting test at 9×10^6 load cycles.

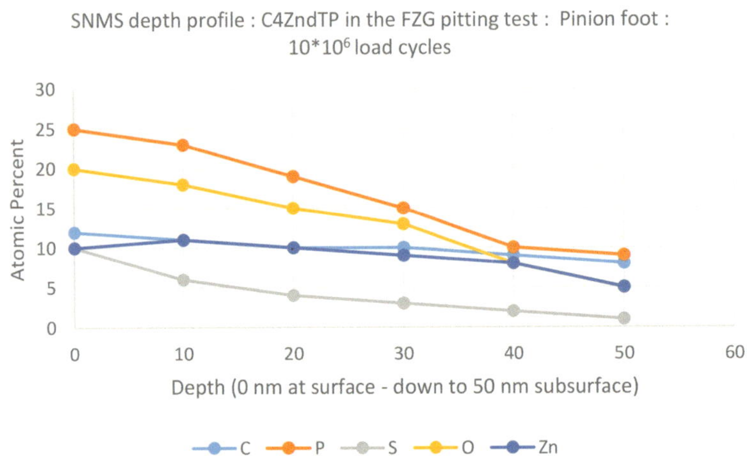

Figure 10.
SNMS depth profile: C4ZndtP in FZG pitting test at 10 × 10[6] load cycles.

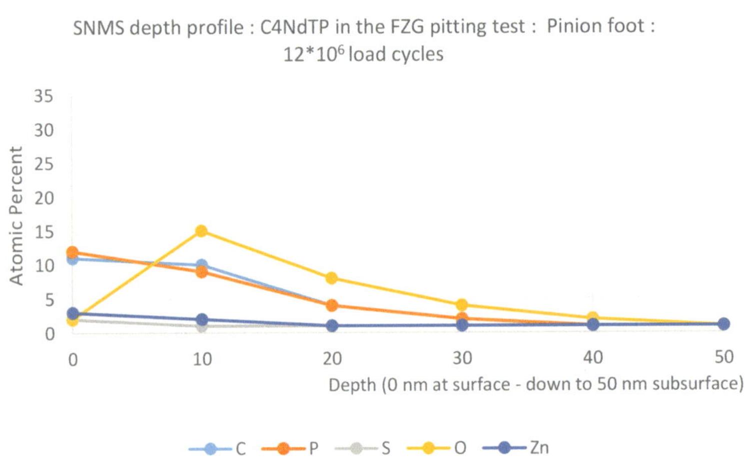

Figure 11.
SNMS depth profile: C4NdtP in FZG pitting test at 12 × 10[6] load cycles.

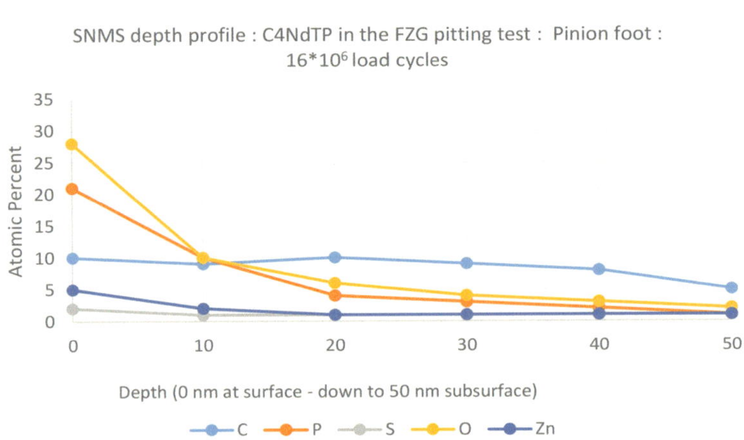

Figure 12.
SNMS depth profile: C4NdtP in FZG pitting test at 16 × 10[6] load cycles.

As a result from *A1–A2*, C4ZndTP causes pitting and increases in layer thickness formation, while *B1–B2* C4NdtP does not at prolong load cycles however shows an increase in surface reaction of the phosphorus component while the reaction layer stays constant.

4. Results for the bearing (NJ406)

The calculation of the load distribution is shown in **Table 5** and **Figure 13**.

The maximum force is acting on roller nr. 7 with a contact pressure of 1481 N/mm^2 [7].

The results (see **Figures 14** and **15**) show an impact of zinc, assumed to be a mixture of phosphates and zinc oxide in the case of the C4ZndtP at 19×10^6 load cycles (**Figure 14**), while compared with the oxygen in the case of the C4NdtP stays low (**Figure 15**).

The bearing thus gives a different reaction by embedding zinc oxide in the near surface. The results for the C4NdtP are quite similar to the reactions seen in the gear.

Speed of the Innerring	2250 m/s
Speed of the Cage	858 m/s
Speed of the Rollers	- 4474 m/s
Tangential Speed at the Inner Ring Raceway	3.28 m/s
Tangential Speed at the Rollers	3.28 m/s
Sum of Speed	6.56 m/s
Pressure at Roller Nr 7 versus Innerring	1481 N/mm^2

Table 5.
Conditions at the bearing NJ406.

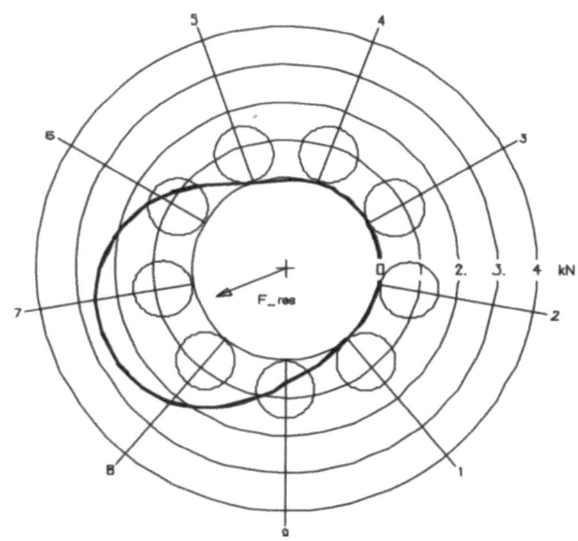

Figure 13.
Load distribution for the NJ406 bearing.

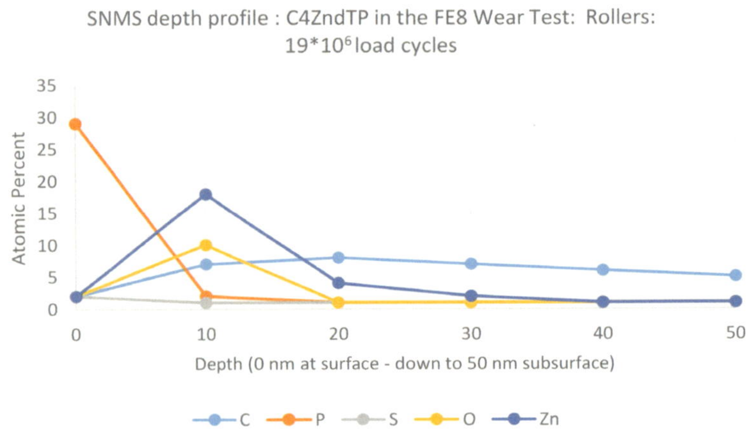

Figure 14.
SNMS depth profile: rollers, C4ZndtP in FE8 bearing test at 19 × 10⁶ load cycles.

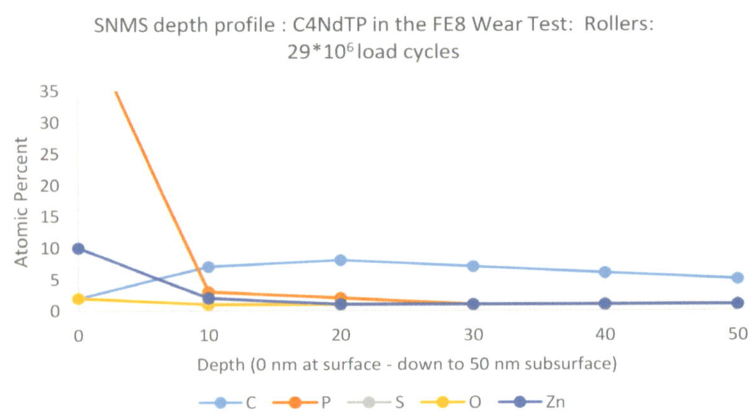

Figure 15.
SNMS depth profile: C4NdtP in the FE8 wear test, rollers at 29 × 10⁶ load cycles.

5. Gear: reaction rates

For the gear (pinion, dedendum) the reaction turnover stays constant or slightly decreases for the C4ZndtP (**Figure 16**) but increases in depth by the use of C4NdTP (**Figure 17**).

The reaction film thickness shows a progression in the case for the C4ZndtP (**Figure 18**), while the C4NdtP shows a regression in time (**Figure 19**).

6. Gear: nanohardness measurement at the pinion

Nanohardness measurements are shown in **Figures 20–22: Figure 20** shows the as-received hardness profile of the dedendum, pitch, and addendum for the as-received pinion tooth flank material (case-hardener SAE 4320).

Figure 21 shows a *steep decrease* by the use of the C4ZndtP compared to the as-received material at the surface.

Figure 22 shows a *steep increase* by the use of C4NdtP compared to the as-received material at the surface.

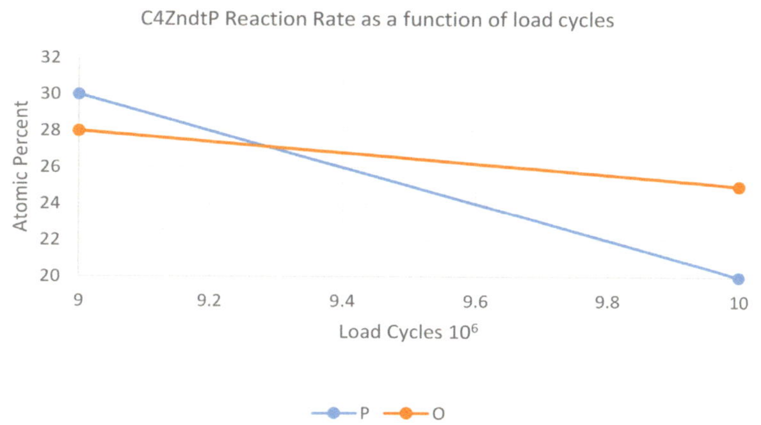

Figure 16.
SNMS profiles: reaction rate (elements phosphorus and oxygen) in the FZG gear tests for C4ZndtP as a function of load cycles.

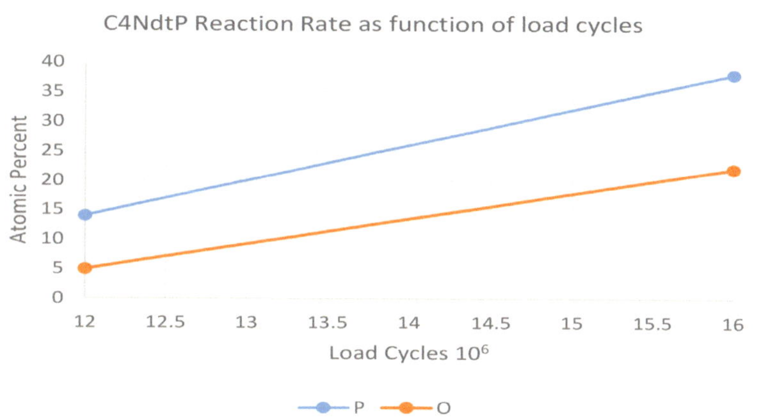

Figure 17.
SNMS profiles: reaction rate (elements phosphorus and oxygen) in the FZG gear test for C4NdtP as a function of load cycles.

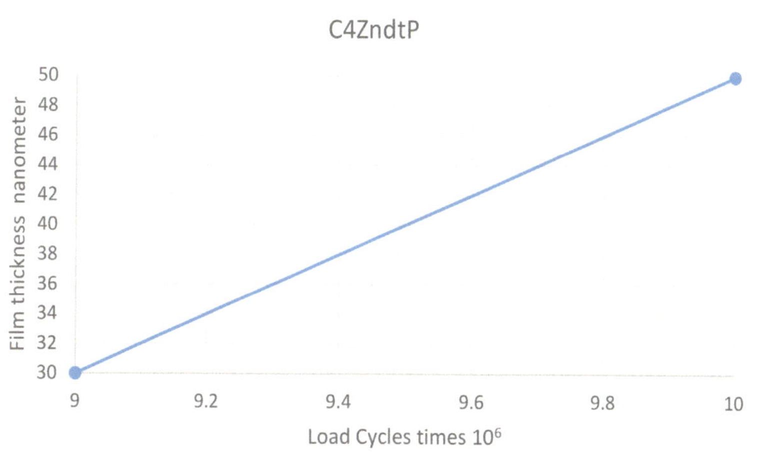

Figure 18.
C4ZndtP: process of film thickness formation as a matter of load cycles.

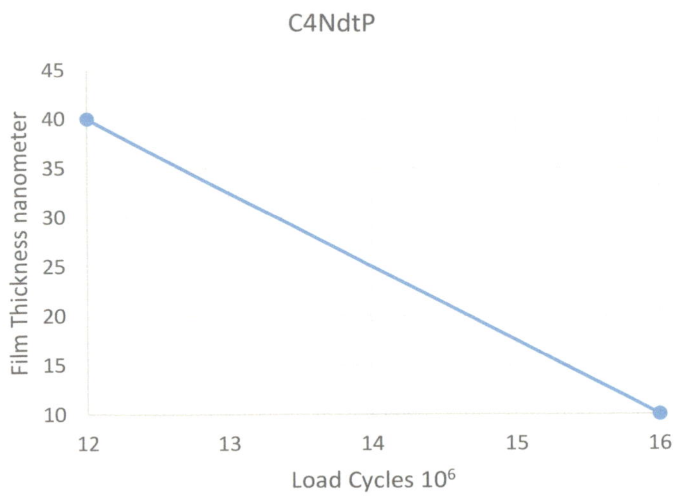

Figure 19.
C4NdtP: process of film thickness formation as a matter of load cycles.

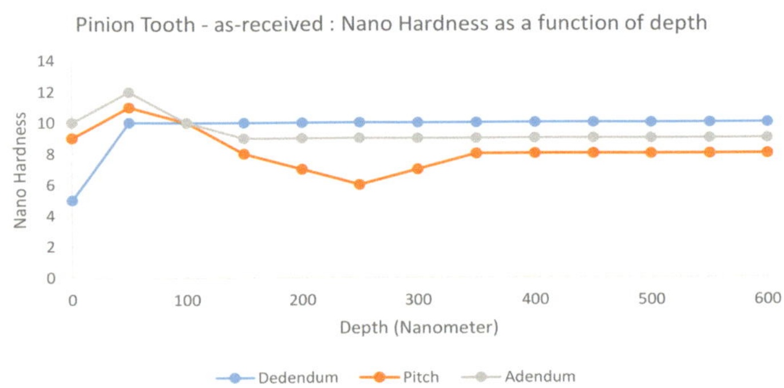

Figure 20.
Nanohardness measurements for the as-received pinion (from dedendum via pitch to the addendum).

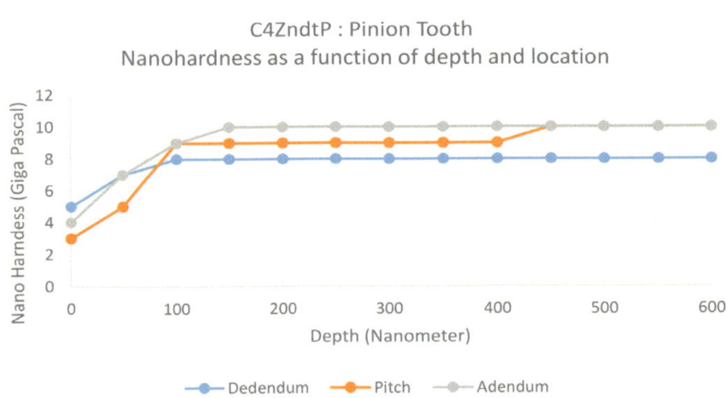

Figure 21.
C4ZndtP: pinion tooth nanohardness as a function of depth (nanometer) and location (dedendum, pitch, and addendum).

7. Molecular description

As functional groups in additives determine the reliability of drivetrain components, it is of interest how those processes are to interpret. Coming from the molecular perspective with a size of 10^{-9} m, it takes effort to interpret effects

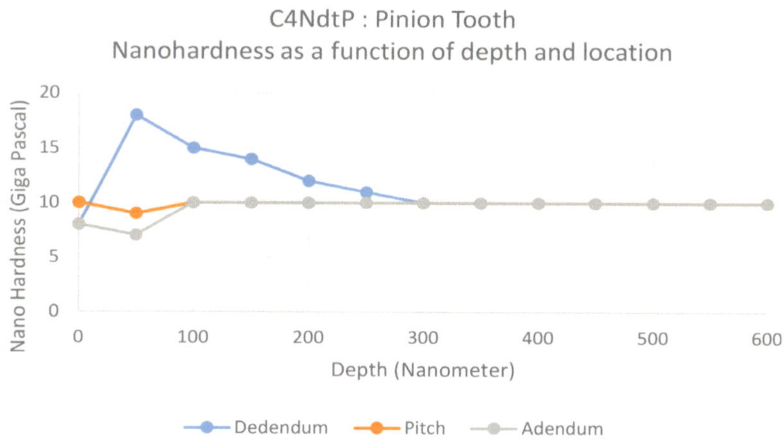

Figure 22.
C4NdtP: pinion tooth nanohardness as a function of depth (nanometer) and location (dedendum, pitch, and addendum).

on 10^{-9} till 10^{-3} m, e.g., magnitudes of 10^6 in length scale. However, considerable progress in multi-scale modeling has become real in the last years; it is of interest how to predict the observed effects reported here by the use of predictors. Basically predicators are obtained by the properties of a molecule, e.g., coming from the chemical bonding. Exploring molecules by quantitative structure property relationship (QSPR) [24] and the molecular properties by the use of density functional theory (DFT) is a standard [25]. The interaction of molecules with themselves and with surfaces is part of molecular dynamics and ab initio methods.

Figure 24A and **B** shows the surface of the additive C4ZndtP with one molecule PAO (as a hydrogenated Di-Dec-1-ene, $C_{20}H_{42}$) (A) and the additive C4NdtP with one molecule PAO (as a hydrogenated Di-Dec-1-ene, $C_{20}H_{42}$) (B) energy minimized by the use of molecular dynamics.

Figure 24A shows the surface of the additive C4ZndtP with one molecule PAO (as a hydrogenated Di-Dec-1-ene, $C_{20}H_{42}$) and the additive C4NdtP with one molecule PAO (as a hydrogenated Di-Dec-1-ene, $C_{20}H_{42}$) (B) attached to an ideal body-centered cubic (bcc) iron surface as C for the C4ZndtP and D for the C4NdtP, energy minimized by the use of molecular dynamics.

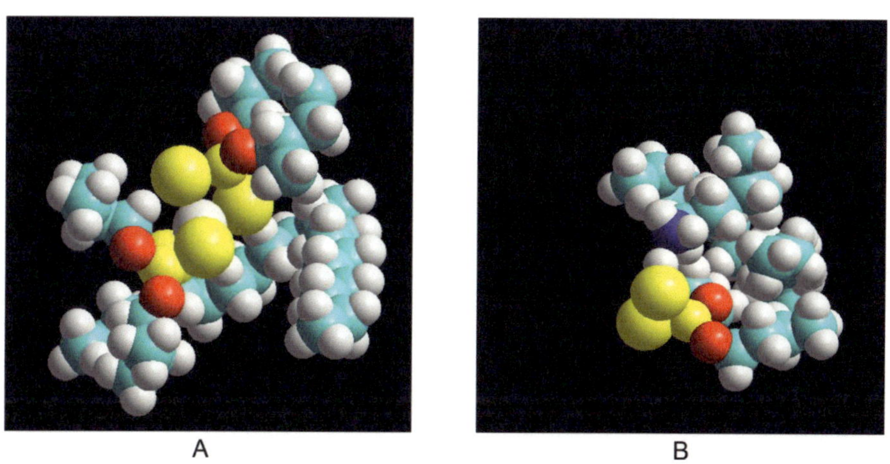

Figure 23.
(A) C4ZndtP structure in PAO and (B) C4NdtP structure in PAO.

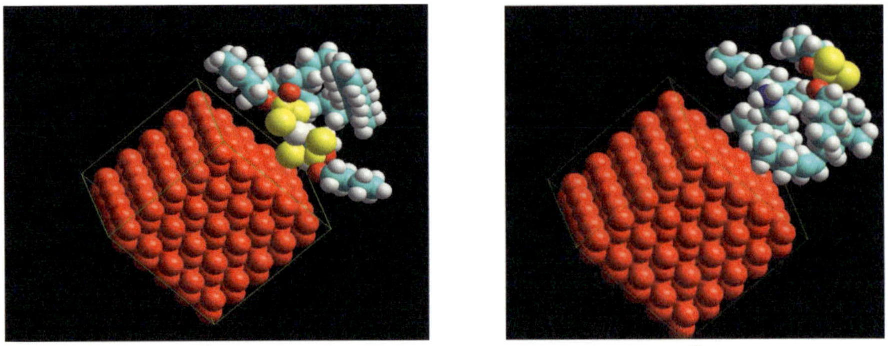

Figure 24.
(A) Approaching a (A) C4ZndtP and (B) C4NdtP to an ideal bcc, iron surface. Labeled atoms are: red: iron; dark red: oxygen; blue: carbon; gray: hydrogen; and yellow: sulfur.

Figure 23B shows the surface of the additive C4ZndtP with one molecule PAO (as a hydrogenated Di-Dec-1-ene, $C_{20}H_{42}$), and the additive C4NdtP with one molecule PAO (as a hydrogenated Di-Dec-1-ene, $C_{20}H_{42}$) (B) attached to an ideal body-centered cubic (bcc) iron surface as C for the C4ZndtP and D for the C4NdtP, energy minimized by the use of molecular dynamics. Approaching this system to an ideal iron surface, it is obvious that the C4Zn is attached with the polar edge (Zn, P, S) to the surface (see **Figure 24A**), while the C4NdtP is attached via the carbon shell (see **Figure 24B**).

8. Discussion

The results shown here may give a reasoning about the elementary analyses found by SNMS where the C4ZndtP progressively acts in time by increasing the reaction layers toward 50 nm constituted by P, O, and Zn oxides, while the C4NdtP shows an initial reaction in the beginning, but regressing the layer to a constant film at 10 nm [26].

As for the C4ZndtP, the reactive core is near to the surface; the reaction may proceed by continuous load cycling, which is found in the SNMS profiles. Due to the continuous degression of the surface toward oxides, the C4ZndtP shows a decrease in the nanohardness by the fact that the surface gets covered with material softer than the base. Also the reaction rate goes down due to fact that the reaction layers are chemically inert compared to iron. The remote position of the reactive group in the C4NdtP exposes the sulfur-phosphorus core to the environment as oxygen. Tribological impacting may then promote the oxidation of the reactive site, rather than a reaction with the metal surface. This means that in the first step the C4NdtP reacts with oxygen at the reactive site, coming to phosphoric acid specie. Those would turn to the surface as they are not soluble in the base oil and naturally get attracted by the oxide sites at the metal surface. The amine would be dissolved back into the base oil. As a fact those phosphoric acid specie are found to be detached on the surface of the pinion dedendum. The oxidation will continue; hence, it is expected that the phosphorus-oxide layer will increase on top, but no material will be leached out due to the fact that the phosphates and polyphosphates are uniquely covering the surface, not being soluble in the matrix.

While the C4ZndtP obviously causes a successive exchange of near-surface material (e.g., iron), the C4NdtP does not. The hardness profiles might be coherent with the carbon profile (SNMS): while the C4ZndtP converts constantly the surface material by smooth oxides, the C4NdtP creates a thin phosphorus-oxide layer on top on a size of 10 nm. The carbon site exposed to the metal might protect it against oxidation, and as the reactive phosphorus-sulfur site is remote, the hardness at least does not go down. The steep increase could be caused by a hardening process of the surface due to carbide formation at the interface as a degradation process of the carbon site. It is noteworthy to say that this interpretation is related to the positions of negative slip and speculative.

Hence, the structure of an additive determines how it approaches and how the subsequent reactions take place, either on the site of the functional head or on the site of the carbon, ending up in the reliability of the application with respect to pitting.

9. Conclusions

The reliability of drivetrain with respect to its expected life cycle is of key interest in the value chain of an installation. Each component contributes to this by the matter of load impacting the load capacity of the materials involved. As load capacity is well defined for the construction materials, e.g., gears and bearings, this definition becomes vague for lubricants. Even though a malfunction of a lubricant could cause damage features, like wear, friction, and tribocorrosion, the understanding of the real function and how to judge it by robust predictors is still missing. Lubricants may give malfunction even in the case of a proper application due to the interaction of functional additives with the mating surfaces. Plenty of contributions worldwide show that the “construction” of a lubricant by adding functional additives into a base oil may lead to premature failures given by the interaction of the functional additives with the given surface. Normally additives are readily dissolved in a base oil and as such transported to the points of interacting surfaces, there getting released in order to uptake a function like wear prevention or friction reduction. However, the energy offered by the contact due to sliding and contact pressure makes additives reactive, causing chemical reactions. The chemical reactions with different additives are seen by the use of specific test conditions, presented in the study as an FZG back-to-back gear test rig. The study brings out that a traditional anti-wear additive such as a zincdithiophosphate (C4ZndtP) reacts continuously at a given threshold with the surface, exchanging the near-surface material. The softening causes continuously material loss over time, ending up in pitting. In contrast, just by changing the chemical structure from a zincdithiophosphate to an ionic liquid like amine-neutralized dithiophosphate (C4NdtP); it is obvious that the application fulfills the complete life cycle without pitting. Compared to the zincdithiophosphate (C4ZndtP), it comes out that the amine-neutralized dithiophosphate (C4NdtP) hardens up at the area of negative slip at the pinion dedendum. Technical data are not to explain this elementary topic. Hence, it has to be seen in a deeper aspect. As additives are dissolved readily in the base oil, the tribological process makes them approach the surface. This brings up the question how the additive is released from the base oil toward the surface as the initial step. In the given example, a simple molecular model shows that in the case of the zincdithiophosphate, the additive approaches the surface with the reactive site given by the sulfur and phosphorus core, continuously leaching iron out of the surface with a subsequent weakening created by reaction layers with little binding to the core of the material. In the case of ammonium-neutralized dithiophosphate, the molecular model shows that this additive approaches the surface by the carbon site, while the sulfur-phosphorus site is remote. This additive gives a hardness increase during the tribological interaction, and as a speculation, the tribological energy may crack the molecule to carbon specie, subsequently hardening the surface up by carbides and preventing an excessive penetration of reaction products.

10. Summary

Additives are part of a drivetrain reliability. It comes clearly that within a construction, the tribological energy offered by the kinematics, the surrounding temperature and environment plus the material involved, has to be judged in terms of the structure of lubricants in the molecular level and how those structures compete with the offer of tribological energy.

Starting from a very basic and standard molecular model, it is essential to understand how additives dissolve in a base oil and how they get released and redissolved at a tribological contact area. Even though how additives act toward a surface might be a minor question, it turns out to be very essential and at least the limiting factor of an application reliability if the criticality of those processes are unknown and might pop up in a given application as premature failure.

Abbreviations

C4ZndtP	isobutyl-zincdithiophosphate
C4NdtP	isobutyl-dithiophosphoric acid reacted with an alkylamine
FZG	gear test rig (DIN ISO 14635)
FE8	bearing test rig (DIN 51819)
PAO	poly-α-olefine as a hydrogenated poly-dec-1-ene
SNMS	secondary neutral mass spectrometry
CRB	cylindrical roller bearing
bcc	body-centered cubic
MPa	megapascal (10^6 Pa)

Author details

Walter Holweger[1,2]

1 Technological Consultant Agency, Epfendorf, Germany

2 Schaeffler Technologies AG & Co. KG, Herzogenaurach, Germany

*Address all correspondence to: walter.holweger@t-online.de

References

[1] Kragelski IV, Dobycin MN, Kombalov VS. Grundlagen der Berechnung von Reibung und Verschleiß. München, Wien: Carl Hanser Verlag; 1982. ISBN: 3-446-13619-3

[2] Bowden T. Friction and Lubrication of Solids. Oxford/Berlin: Clarendon Press/Springer; 1950. ISBN: 978-3-642-92755-3

[3] Zum Gahr K-H. Microstructure and Wear of Materials. Amsterdam: Elsevier; 1987. ISBN: 0-444-42754-6

[4] Rabinowitcz E. Friction and Wear of Materials. New York: Wiley-Interscience; 1995. ISB-0-471-83084-4

[5] Czichos H, Habig K-H. Tribologie-Handbuch. Wiesbaden: Vieweg & Teubner; 2010. ISBN: 978-3-83480-017-6

[6] Bushan B. Modern Tribology Handbook, Vol. 1(2). New York: CRC Press; 2001

[7] FVA Report, 289 I & II. Zusammenhänge zwischen Zahnrad- und Wälzlagerschäden und tribologischen Veränderungen des oberflächennahen Werkstoffbereichs; 2001–2003

[8] Johanssen J. On the Influence of Gear Oil Properties on Pitting Life. Lulea: University of Lulea; 2015

[9] FZG Test Rig. Available from: https://www.researchgate.net/publication/235308055_Testing_procedures_for_gear_lubricants_with_the_FZG_test_rig

[10] Geheeb N, Franke J. Forschungsvereinigung Antriebstechnik. FVA 126. RWTH Aachen: Vogel Verlag; 2000

[11] Reichelt M. Mikroanalytische Klärung Des Verschleißschutzes in Langsam Laufenden Wälzlagern. Aachen: Shaker; 2011. ISBN: 978-3-8440-0424-3; ISSN: 1618-5722

[12] Höhn BR, Oster P. Influence of the Lubricants on Pitting and Micro Pitting in the FZG Gear Test Rig. Hamburg: Deutsche Wissenschaftliche Gesellschaft für Erdöl, Erdgas und Kohle e.V.; 1996

[13] Winter H, Michaelis K. AGMA Technical Paper P291.17, scoring load capacity of gears lubricated with EP-oils. In: Fall Technical Meeting; October 17–19; Montreal, Canada; 1983

[14] Schönnenbeck G et al. Mineralöltechnik. 1987;**32**(6):1-24

[15] Vinogradova IE. Antianrißinhibitoren für Öle. Moskau: Verlag Chimija; 1972. p. 272

[16] Ruina A, Pratap R. Introduction to Statics and Dynamics. Oxford: Oxford University Press; 2002. p. 713

[17] Arun AP, Senthil AP, Giriaj B, Faizur RA. Gear test rig—A review. International Journal of Mechanical & Mechatronics Engineering. 2014;**14**. No: 05 16 140205-9696-IJMME-IJENS ©

[18] Hibbeler RC. Engineering Mechanics. 11th ed. New Jersey: Pearson/Prentice Hall; 2007. p. 393. ISBN: 0-13-127146-6

[19] Castro J, Sottomayor A, Seabra J. Experimentals and analytical scuffing criteria for FZG gears. Tribology Series. 2003;**43**:651-661

[20] Fernandes C, Blazquez L. FZG gearboxes lubricated with different formulations of polyalphaolefin wind

turbine gear box oils. In: International Gear Conference, Lyon; 2014

[21] Jede RH, Peters RH, et al. Analyse dünner Schichten mittels Massenspektrometrie zerstäubter Neutralteilchen. TM—Technisches Messen. 1986;**11**:407-413

[22] Passlack S, Kopnarski M. Sekundärneutralteilchen-Massenspektrometrie (SNMS). 2014. Available from: https://onlinelibrary.wiley.com/doi/pdf/10.1002/vipr.201400569

[23] Vad K, Csik A, Langer G. Secondary neutral mass spectrometry—A powerful technique for quantitative elemental and depth profiling analyses of nanostructures. Chichester, UK: Spectroscopy Europe; 2009:13-16. Available from: https://www.spectroscopyeurope.com/article/secondary-neutral-mass-spectrometry-powerful-technique-quantitative-elemental-and-depth

[24] Katritzky AR, Lobanov VS, Karelson M. QSPR: The correlation and quantitative prediction of chemical and physical properties from structure. Chemical Society Reviews. 1995;**24**:279-287. Available from: https://pubs.rsc.org/en/content/articlelanding/1995/cs/cs9952400279

[25] Bockstedte M, Kley A, Neugebauer J, Scheffler M. Density-functional theory calculations for poly-atomic systems: Electronic structure, static and elastic properties and ab initio molecular dynamics. In: Computer Physics and Communications. Vol. 107. Amsterdam: Elsevier; pp. 187-222. Available from: https://www.sciencedirect.com/science/article/pii/S0010465597001173

[26] Szlufarska I. Multi-scale modeling of friction and wear. In: 22nd Conference of Wear of Materials; Long Beach, California; 2018. University of Wisconsin, Key Note

Chapter 6

Tribological and Wear Behavior of Metal Alloys Produced by Laser Powder Bed Fusion (LPBF)

Massimo Lorusso

Abstract

Laser powder bed fusion (LPBF) is an additive manufacturing technique for the production of parts with complex geometry, and it is especially appropriate for structural applications in aircraft and automotive industries. Wear is the most important cause of malfunction of mechanical systems. Abrasive wear accounts for 50% of wear in industrial situations, and it is most common in components of machines. LPBF is very attractive due to its extremely high melting and solidification rates that make possible to obtain materials with particular tribological and wear behavior than those by traditional manufacturing routes. The aim of this chapter is to investigate the different behaviors of principal metallic alloys by LPBF.

Keywords: additive manufacturing (AM), laser powder bed fusion (LPBF), metallic alloys, wear

1. Introduction

According to ASTM F2792-10, additive manufacturing (AM) is defined as "The process of joining materials to make objects from 3D model data, usually layer upon layer, as opposed to subtractive manufacturing technologies." The fundamental principle of AM is that a geometric representation, originally generated using 3D-CAD system, can be manufactured directly without a need to process planning [1].

Today AM is receiving a very high attention from the mainstream media, investment community, national governments, and scientific communities. Nearly 10 years ago (2008), only 231 articles were published with AM topic, 5 years ago (2013) about 800 articles, and in 2018 about 4900 articles; in 10 years the number of articles per year is increased more than 20 times (**Figure 1**).

AM technologies have a strong potential to change the characteristic of manufacturing process, away from mass production in large factories with dedicated tooling and with high costs, to a world of mass customization and distributed manufacture.

Everyday new and innovative applications are emerging for the additive manufacturing [2]:

- Prototyping
- Art and jewelry
- Tooling

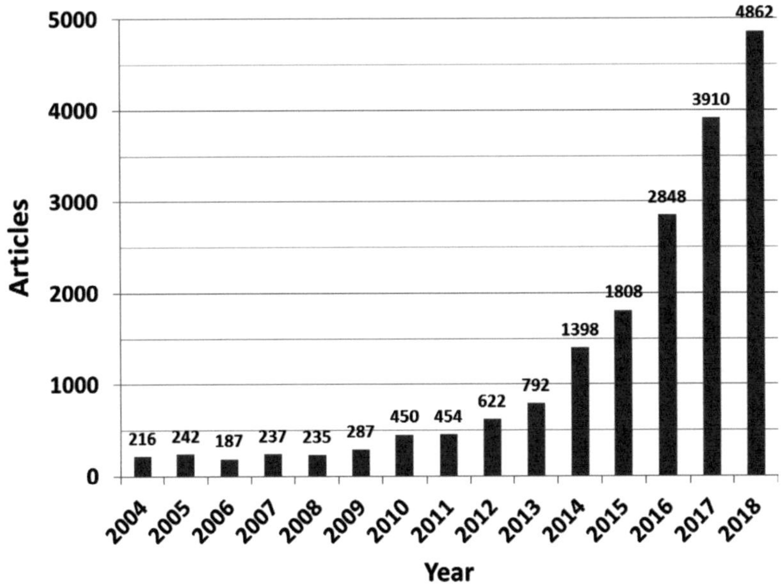

Figure 1.
Number of scientific articles per years with AM topic (source: Scopus).

- Medical and dental
- Automotive
- Aerospace

In many application of the AM, the wear resistance is important to guarantee the efficacy and the safety. Wear is the most important cause of malfunction of mechanical systems; for this reason, it is important to study the effect of wear and generally, the tribological characteristic of material used and processed by AM [3].

The aim of this chapter is to investigate the different behaviors of principal metallic alloys processed by laser powder bed fusion (one of the most diffused AM technologies) in terms of tribological properties, with a particular focus on the wear resistance and the coefficient of friction (COF). At the moment, few studies are available about tribological properties of metallic alloys produced by LPBF; this chapter searches to organize the works present.

2. Additive manufacturing

2.1 Introduction

Seven different technologies are classified for additive manufacturing agreed by the AM SIG (special interest group) as can be seen in detail in **Table 1** [4].

2.2 Laser powder bed fusion (LPBF)

The SLM process has been defined as the laser powder bed fusion process (LPBF), according to ISO/ASTM 52900. It is also known by the trade names LaserCUSING or DMLS (Direct Metal Laser Sintering), which directly produces homogenous metal objects, layer by layer, from 3D CAD data, by selectively melting very fine layers of metal powder (**Figure 2**) with a laser beam.

Classification	Description	Technology	Materials
Direct energy deposition	Builds parts using focused thermal energy and wire to fuse materials and they are deposited on a substrate	Laser deposition Laser consolidation Direct metal deposition Electron beam direct melting	Metals
Binder jetting	Creates objects by deposition of a binding agent to join powdered material	3D printing Ink-jetting S-print M-print	Metals, polymers, and ceramics
Material extrusion	Fused deposition modeling	Fused deposition modeling	Polymers
Material jetting	Builds parts by depositing small droplets of build material, which are then cured by exposure to light	Polyjet Ink-jetting Thermojet	Photopolymers, wax
Powder bed fusion	Creates objects by using thermal energy to fuse regions of a powder bed	Selective laser melting Laser powder bed fusion Selective laser sintering Electron beam melting	Metals, polymers, and ceramics
Sheet lamination	Builds parts by trimming sheets of material and binding them together in layers	Ultrasonic consolidation Laminated object manufacturing	Metals, ceramics, and hybrids
VAT photopolymerization	Builds parts by using a vat of liquid photopolymer resin, out of which the model is constructed layer by layer. An ultraviolet (UV) light is used to cure or harden the resin where required.	Stereolithography Digital light processing	Photopolymers and ceramics

Table 1.
Classification of AM adapted from ASTM AM classification.

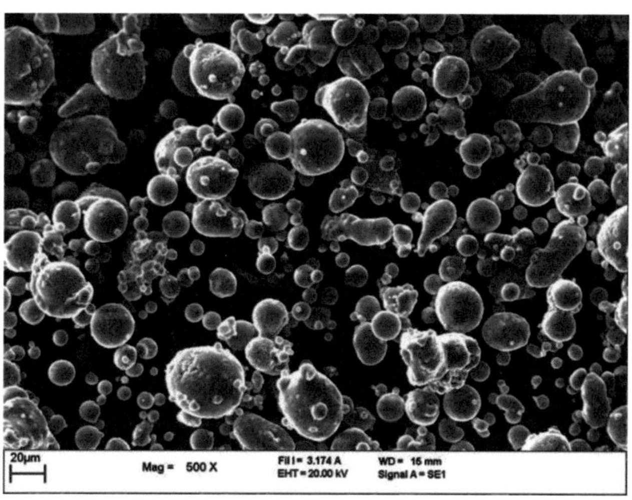

Figure 2.
Scanning electron magnification (SEM) observation of typical powder used in LPBF process ($AlSi_{10}Mg$).

Laser powder bed fusion (LPBF) is an additive manufacturing technique for the fabrication of near net-shape parts directly from computer-aided design data by melting together different layers with the help of a laser source. LPBF process produces parts with good surface quality, high accuracy and detail resolution, and excellent mechanical properties. The components are built layer by layer; it is possible to project internal channel and features that are impossible to obtain by casting or machining. LPBF does not require special tooling like casting, so it is more convenient for not so big production. It is a good alternative to conventional machining for complex metallic parts [5].

It has been demonstrated, in recent literature, that SLM can also be used to fabricate metal matrix composites (MMCs). These could have applications in automotive and aerospace industries, where it is necessary to improve mechanical properties (stiffness and hardness) and specially the wear resistance [6].

3. Aluminum alloys

3.1 Introduction

The most used Al alloys are Al-Si alloys, which represent 80% of aluminum casting alloys, thanks to their high fluidity, high weldability, good corrosion resistance, and low coefficient of thermal expansion. The binary Al-Si system is a eutectic alloy when the amount of Si is 11–13 wt%, a hypoeutectic alloy when Si is less than 11 wt%, and a hypereutectic alloy when Si is more than 13 wt%. The strengthening of these alloys is generally possible, through the addition of other alloying elements such as Cu and Mg that make the Al-Si alloys hardenable either by means of a heat treatment. There is a large demand for Al-Si-Mg alloys for different applications, such as the aerospace industry, and for automotive and heat exchangers, due to their high mechanical properties, like strength and hardness, in the heat-treated state [7–9].

The most popular Al-Si alloy processed by LPBF is $AlSi_{10}Mg$ alloy (similar to A360). Other Al-Si alloys by LPBF are $AlSi_7Mg$ (called also A357) [10–11] and $AlSi_{12}Mg$ [12].

Despite this growing interest in the AM processability of Al-Mg-Zn-Cu alloys, to date, few studies are available on the AM process of high mechanical properties' (harness and strength) aluminum alloys. It is well-known that the alloys belonging to the Al-Mg-Zn-Cu alloys are appropriate for different applications in aerospace as they are characterized by toughness and high strength reached mainly through the precipitation of the $MgZn_2$ phase. These alloys are not well weldable because they suffer strongly from liquation cracking [13–14].

3.2 $AlSi_{10}Mg$

In the literature, it is demonstrated that $AlSi_{10}Mg$ alloy produced by casting has a coefficient of friction (against a WC cemented with CO pin) lower than $AlSi_{10}Mg$ alloy by LPBF since their microstructure and hardness are different. The typical microstructure of metallic alloys by LPBF without heat treatments is characterized of a small grain size. At higher magnification after hatching, it can be seen as a fine cellular-dendritic structure made by agglomerates of grains with mean diameters of a hundred of nanometers or less. It is generally observed that materials with large grains have a COF lower than materials with a fine microstructure; this is one of the most important reasons of higher COF of $AlSi_{10}Mg$ by LPBF [15].

The different sizes of microstructure influence the hardness very strong. The hardness is higher for the finer grain size. As suggested by the theoretical considerations, the material with the highest hardness has the highest wear resistance. The difference between the wear resistance of the $AlSi_{10}Mg$ alloy produced by casting and by LPBF is immediately evident. During pin on disc test, the volume per meter loss of the $AlSi_{10}Mg$ produced by LPBF is 35% less than the volume per meter loss of the $AlSi_{10}Mg$ produced by casting.

3.3 Other aluminum alloys

In general, for the conventional casted alloys, the Al-Si alloys with small primary silicon phase present a higher wear resistance than that of the alloys with large silicon phase, due to their high surface-volume fraction. The aluminum alloys by LPBF show the inverse results that could be attributed to their ultrafine microstructure.

During the wear process, the fine primary silicon particles form a full contacted wear layer; the primary silicon is directly pressed into Al-matrix and then forms the full contacted wear layer [7]. For this reason alloys with small primary silicon have a relative poor wear resistance. The A357 aluminum alloy has less silicon (6.5–7.56%) than $AlSi_{10}Mg$ (9–11%) but higher COF and wear [16].

For the Al-Zn-Mg alloys, the microstructure has a strong influence on the wear behavior that is due to higher content and the higher amount of $MgZn_2$ precipitate that is harder than α-aluminum matrix and helps to protect the surface of material [13].

3.4 Aluminum matrix composites (AMCs)

Aluminum matrix composites (AMCs) have generally excellent mechanical properties such as improved stiffness, strength, and hardness when compared with the aluminum matrix. AMCs attract much attention because they are characterized by low density and high specific strength and good tribology properties. The limits of this material are the high difficulty in the process of production and in the post-processing phases. The principal problem when a ceramic is used as reinforcement is the clustering and agglomeration caused by the poor wettability and a large surface-to-volume ratio that does not promote a homogenous dispersion.

The LPBF process seems to be particularly suitable for the production of AMCs because near net-shape complex components can be made, which reduces the post-processing phases. The most used ceramic reinforcements in AMCs produced by LBPF are magnesium spinel ($MgAl_2O_4$) and titanium diboride (TiB_2) [15].

The sufficiently high densification rate combined with the homogeneous incorporation of nanoscale TiC reinforcement throughout the matrix led to the considerably low coefficient of friction (COF) and resultant wear rate [17].

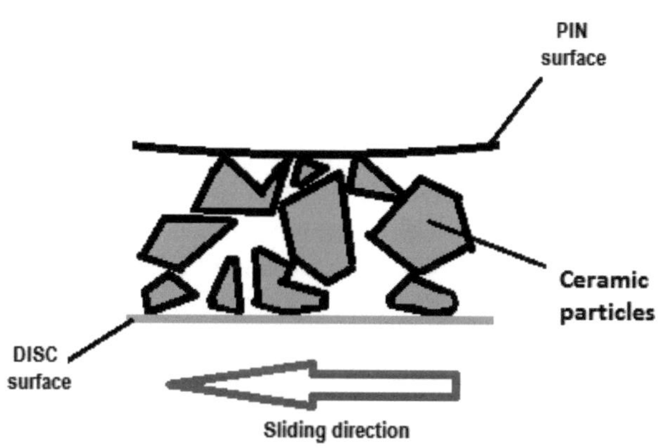

Figure 3.
Example of detached particles that have an effect of solid lubrication (third part).

The presence of reinforcements causes a reduction of COF. If the reinforcements have a micro-size, the effect is bigger than with nano-sized reinforcement. The reduction of COF is probably due to the detachment from the aluminum matrix of micro- or nanoparticles of ceramic reinforcements that can act as a third body (**Figure 3**).

4. Nickel alloys

4.1 Introduction

The most used nickel alloys produced through LPBF for their high weldability are Inconel 718 and 625. Inconel 718 and 625 have been used in high-temperature applications, such as nuclear reactors, pumps, molds, and gas-turbine engine aircraft. These nickel alloys are endowed with high-temperature strength, high creep, and oxidation resistance. These two alloys can be used depending on the applications, but the production of objects with a complex shape is expensive with conventional manufacturing technologies. Therefore, the ability to produce complex components without using molds makes LPBF process particularly interesting. The microstructure created during LPBF process are out-of-equilibrium, and it is necessary to perform some heat treatments in order to homogenize the microstructural features. Depending to application it would be necessary to carry out a simple stress relieving to reduce residual stress induced by thermal gradients during LPBF process. As for other applications, annealing or solution treatment to allow the grain recrystallization and growth, thus improving the creep resistance, is typically requested [18–21].

4.2 Inconel

Inconel 625 is a nickel-chromium alloy designed as solid-solution-strengthened. The heat treatments favor the formation of metastable Y" phase that further improves the mechanical properties. Inconel 625 thanks to higher concentration of Cr and Mo with respect to Inconel 718 has higher corrosion resistance.

Inconel 718 is an age-hardenable nickel-chromium alloy mainly due to the presence of aluminum, titanium, and niobium that leads to precipitation of gamma prime Y' $Ni_3(Al,Ti)$ phase and metastable gamma double prime Y" Ni_3Nb phase [18]. Inconel by LPBF is more difficult to be machined than the same materials produced by extruding or rolling processes. During milling process of Inconel by LPBF, the cutting speed and chip load are lower due the presence of hard precipitated particles. In general, the Inconel produce by LPBF exhibited relatively good wear performance. Such microstructures indicated that the presence of severe adhesive wear in turn resulted a relatively higher wear rate. The clustered Y dendrites gave rise to the fluctuations of COF. The formed protective adherent tribolayer on worn surfaces made considerable contributions to the further improved wear performance of LPBF-produced parts. The combined influence of elevated microhardness and the formation of adherent tribolayer contributed to the improvement of wear performance [20].

Different studies, in particular about Inconel 718, show that addition of tungsten carbide (WC) or titanium carbide (TiC) particles significantly increased the hardness, friction resistance, and wear performance. The composite acquired a considerably low COF. The existence of a gradient interface has a very important role in improving the wear performance of LPBF-processed WC/Inconel 718 and TiC/Inconel 718 composites [22].

5. Titanium

5.1 Introduction

Titanium and its alloys have good mechanical properties, good corrosion resistance, and excellent biocompatibility. These alloys are the most interesting metallic biomaterials for orthopedic and dental implants. Until few years ago, titanium processing via AM technologies was given little consideration by the medical industry due to the high cost of production. However, in recent years, AM metal technologies are becoming popular in biomedical field because of the ability to build metals with customized porous architectures and shape. The titanium alloy Ti_6Al_4V (the most popular titanium alloy) has been widely used in various industrial applications due to its mechanical and physical properties. Beyond the biomedical field, Ti_6Al_4V has been commonly employed in producing aircraft engine airframe parts owing to its high strength to mass ratio and good performance at high temperature (up to 400–500°C) [23–24].

5.2 Ti_6Al_4V

Ti_6Al_4V has very good mechanical properties, but it has also been reported to exhibit poor tribology properties, such as a high COF and low wear resistance. The poor tribological property of Ti_6Al_4V is attributable to its low resistance to plastic shearing, low work hardening, and the low protection afforded by surface oxidation. No significant differences are present between TI_6Al_4V produced by the different processing technologies. Generally on Ti_6Al_4V produced by LPBF, less oxidized areas are found during the wear tests [25].

During the LPBF process, the high cooling rate of laser melting leads to higher amount of α and α' harder phases on Ti_6Al_4V alloy than the traditional process. The presence of harder microstructural constituents on Ti_6Al_4V produced by LPBF leads to a higher wear resistance. The heat-treatment Ti_6Al_4V generates a protective tribolayer containing oxygen without plastic deformation in the bulk material, which has the lowest wear rate [26].

Investigation of reinforced Ti_6Al_4V with TiB_2 shows that nano-sized TiB whiskers are formed by the in situ reaction between Ti and TiB_2. The interface between matrix and TiB is a very strong interface bonding. During the wear test, this avoids the possibility of easy detachment of TiB whiskers. This reduces the wear rate significantly but not the COF, because the detached particles are few and it is not present enough third part that reduces significantly the friction [27].

6. Stainless steel (316L)

The 316L austenitic stainless steel has numerous application in different fields for its high resistance at oxidation and corrosion. The most popular applications are in marine, nuclear, oil and gas, and biomedical industry. 316L austenitic stainless steel which comprises iron alloyed with chromium of mass fraction up to 18%, nickel up to 14%, molybdenum up to 3% , manganese down 2%, silicon down 0.75%, copper down 0.5% and carbon down 0.03% along with minor elements [28].

During wear test on 316L stainless steel produced by LPBF, the passive layer made by chromium and nickel oxidation is removed and leaves iron exposed to the air, which easily gets oxidized especially at high temperature. Regarding the wear mechanisms, the worn surfaces of 316L stainless steel exhibited plastic deformation due to adhesive wear as well as grooves aligned along the sliding direction due to the abrasive wear. The wear rate and the friction of 316L stainless steel by LPBF were lower than the 316L traditionally processed; the LPBF-processed steel has a very fine austenite grains, the size of which was much smaller than in the traditional-processed 316L stainless steel. These fine grains in the 316L stainless steel by LPBF increase the wear resistance, and the surface is subjected to slight plastic deformation [29–30].

Investigation of the wear resistance of reinforced 316L stainless steel (with TiB_2 or TiC) shows that the wear resistance increases with the increasing TiB_2 content due to combined effects of grain refinement and grain-boundary strengthening [31].

7. Lubrication condition and heat treatment

Metallic alloys by LPBF have generally pores and cracks that influence the wear under lubricated condition. Few studies are available in literature under boundary lubrication regime.

In those studies [26, 30], when a lubricating film is not yet formed, the metal alloys by LPBF have better wear performance than metal alloys by traditional processes.

Surface pores may positively influence the formation of the lubricating film.

The effect of lubricant is critical in reducing friction and wear. The choice of oil needs to be carefully considered before applying LPBF process to hydraulic components.

In general, heat treatment (used to reduce the stress in material after LPBF process) reduces the wear resistance of metallic alloys. The wear resistance is reduced because the heat treatment changes the microstructure of metallic alloys by LPBF and lost the very fine microstructure. The most sensibility materials at heat treatment are aluminum alloys [15, 32]. The only metallic alloy that increases the wear resistance and reduces COF after heat treatments is Ti_6Al_4V because the oxidation of surface (if the heat treatment is realized in the presence of oxygen) produces a protective tribo-oxide layer [26].

8. Conclusions

In conclusion, metallic alloys by LPBF generally have higher wear resistance and less COF than metallic alloys produced by traditional processes under dry condition and boundary lubrication mainly due to the fine grains and high hardness.

The LPBF processing parameters are fundamental for wear rate since a fully densified part usually has high wear resistance and COF.

The existence of pores reduces the bonding between molten pools, resulting in cracks. These cracks can further cause material shell off which greatly increases wear.

In general, the metallic alloys produced by LPBF are more difficult to machine than the same metallic alloys produced by traditional processes. For this, it is important to reduce at the minimum the post-process machining.

The presence of ceramic particle reinforcements in MMCs causes generally a reduction of COF; this effect is due to the detachment from the metallic matrix of ceramic particles that can act as a third body. The interfacial bond between the matrix and the reinforcements has a fundamental role in wear process; a strong interfacial bond guaranties a low wear rate; a weak interfacial bond causes a low COF and sometimes high wear rate.

The heat treatment in general reduces the wear resistance and increases the COF.

Acknowledgements

The author is grateful for the moral and scientific support of PhD Diego Manfredi, PhD Flaviana Calignano, and Professor Matteo Pavese.

Author details

Massimo Lorusso
Istituto Italiano di Tecnologia (IIT), Torino, Italy

*Address all correspondence to: massimo.lorusso@iit.it

References

[1] Gibson I, Rosen WD, Stucker B. Additive Manufacturing Technologies-Rapid Prototyping to Direct Digital Manufacturing. New York, USA: Springer; 2010

[2] Wohlers T. Wohlers Report 2017 State of Industry, Annual Worldwide Profess Report. Fort Collins CO, USA: Wohler Associates; 2017

[3] Cordovilla CG, Narciso N, Louis E. Abrasive wear resistance of aluminum alloy/ceramic particulate composites. Wear. 1996;**192**:170-177. DOI: 10.1016/0043-1648(95)06801-5

[4] Manfredi D, Calignano F, Krishnan M, Canali R, Ambrosio EP, Biamino S, et al. Additive Manufacturing of Al Alloys and Aluminium Matrix Composites (AMCs) in Light Metal Alloys Applications. InTech; 2014. pp. 3-34. DOI: 10.5772/58534

[5] Calignano F, Manfredi D, Ambrosio EP, Iuliano L, Fino P. Influence of process parameters on surface roughness of aluminum parts produced by DMLS. International Journal of Advanced Manufacturing Technology. 2013;**67**:2743-2751. DOI: 10.1007/s00170-012-4688-9

[6] Aversa A, Marchese G, Lorusso M, Calignano F, Biamino S, Ambrosio EP, et al. Microstructural and mechanical characterization of aluminum matrix composites produced by laser powder bed fusion. Advanced Engineering Materials. 2017;**19**:11. DOI: 10.1002/adem.201700180

[7] Kang N, Coddet P, Chen C, Wang Y, Liao H. Coddet C; Microstructure and wear behavior of in-situ hypereutectic Al-high Si alloys produced by selective laser melting. Materials and Design. 2016;**99**:120-126. DOI: 10.1216/j.matdes.2016.03.053

[8] Brito C, Reinhart G, Nguyen-Thi H, Mangelinck-Noel N, Cheung N, Spinelli JE, et al. High cooling rate cells, dendrites, microstructural spacings and microhardness in a directionally solidified Al-Mg-Si alloy. Journal of Alloys and Compounds. 2015;**636**:145-149. DOI: 10.1016/j.jallcom.2015.02.140

[9] Trevisan F, Calignano F, Lorusso M, Pakkanen J, Aversa A, Amborsio EP, et al. On the selective laser melting (SLM) of the $ALSi_{10}Mg$ alloy: Process, microstructure and mechanical properties. Materials. 2017;**10**:76. DOI: 10.3390/ma10010076

[10] Aversa A, Lorusso M, Trevisan F, Amborsio EP, Calignano F, Manfredi D, et al. Effect of process and post-process conditions on the mechanical properties of an A357 alloy produced via laser powder bed fusion. Metals. 2017;7:68. DOI: 10.3390/met7020068

[11] Rao H, Giet S, Yang K, Wu X, Davies CHJ. The influence of processing parameters on aluminium alloy A357 manufactured by Selective Laser Melting. Materials and Design. 2016;**109**:334-346. DOI: 10.1016/j.matdes.2016.07.009

[12] Leary M, Mazur M, Elambasseril J, McMillan M, Chirent T, Sun Y, et al. Selective laser melting (SLM) of $AlSi_{12}Mg$ lattice structure. Materials & Design. 2016;**98**:344-357. DOI: 10.1016/j.matdes.2016.02.127

[13] Reis BP, Lopes MM, Amauri G, Dos Santos CA. The Correlation of microstructure features, dry sliding wear behavior hardness and tensile properties of AL-2wt% Mg-Zn alloys. Journal of Alloys and Compounds. 2018;**764**:267-278. DOI: 10-1016/j.jallcom.2018.06.075

[14] Aversa A, Marchese G, Manfred D, Lorusso M, Calignano F, Biamino S, et al. Laser powder bed fusion of a high

strength Al-Si-Zn-Mg-Cu alloy. Metals. 2018;**8**:300. DOI: 10.3390/met8050300

[15] Lorusso M, Aversa A, Manfredi D, Calignano F, Ambrosio EP, Ugues D, et al. Tribological behavior of aluminum ally $AlSi_{10}Mg$-TiB_2 composites produced by direct metal laser sintering (DMLS). Journal of Materials Engineering and Performance. 2016;**25**:3152-3160. DOI: 10.1007/s11665-016-2190-5

[16] Lorusso M, Trevisan F, Aversa A, Calignao F, Amborsio EP, Pavese M, et al. Macro-, micro- and nano-hardness and macro- and nano- wear behavior of aluminum alloys by laser powder bed fusion. In: Proceeding of the European Congress and Exhibition on Advanced Materials and Process (EUROMAT2017); Thessaloniki, Greece. 2017

[17] Gu D. Nanoscale TiC particle-reinforced $AlSi_{10}Mg$ bulk-form nanocomposites by selective laser melting (SLM) additive manufacturing (AM): Tailored microstructures and enhanced properties. In: Laser Additive Manufacturing of High-Performance Materials. Berlin: Springer; 2015. pp. 175-199. DOI: 10.1007/978-3-662-46089-4_6

[18] Marchese G, Bassini E, Calandri M, Ambrosio EP, Calignano F, Lorusso M, et al. Microstructural investigation of as fabricated and heat-treated Inconel 625 and Inconel 718 fabricated by direct metal laser sintering: Contribution of Politecnico di Torino and Istituto Italiano di Tecnologia (IIT) di Torino. Metal Powder Report. 2016;**71**:273-278. DOI: 10.1016/j.mprp.2016.06.002

[19] Marchese G, Colera XG, Calignano F, Lorusso M, Biamino S, Minetola P, et al. Characterization and comparison of Inconel 625 processed by selective laser melting and laser metal deposition. Advanced Engineering Materials. 2016;**19**:3. DOI: 10.1002/adem.201600635

[20] Jia Q, Gu D. Selective laser melting additive manufacturing of Inconel 718 superalloy parts: Densification, microstructure and properties. Journal of Alloys and Compounds. 2014;**585**:712-721. DOI: 10.1016/j.jallcom.2013.091.171

[21] Zhang DY, Niu W, Cao XY, Liu Z. Effect of standard heat treatment on the microstructure and mechanical properties of selective laser melting manufactured Inconel 718 superalloys. Materials Science and Engineering A. 2015;**644**:32-40. DOI: 10.1016/j.msea.2015.06.021

[22] Rong T, Gu D, Shi Q, Cao S, Xia M. Effects of tailored gradient interface on wear properties of WC/Inconel 718 composites using selective laser melting. Surface & Coating Technology. 2016;**307**:418-427. DOI: 10.1016/j.surfcoat.2016.09.011

[23] Trevisan F, Calignano F, Aversa A, Marchese G, Lombardi M, Biamino S, et al. Additive manufacturing of titanium alloys in the biomedical properties and applications. Journal of Applied Biomaterials and Functional Materials. 2018;**16**(2):57-65. DOI: 10.5301/jabfm.5000371

[24] Bartolomeua F, Sampaioa M, Carvalhoa O, Pintob E, Alvesb N, Gomesa JR, et al. Tribological behaviour of Ti_6Al_4V cellular structures produced by selective laser melting. Journal of the Mechanical Behavior of Biomedical Materials. 2017;**69**:128-134. DOI: 10.1016/j.jmbbm.2017.01.004

[25] Bartolomeu F, Buciumeanu M, Pinto E, Alves N, Silva FS, Carvalho O, et al. Wear behavior of Ti_6Al_4V biomedical alloys processed by selective laser melting, hot pressing and conventional casting. Transactions of the Nonferrous Metals Society of China. 2017;**27**:829-838. DOI: 10.1016/S1003-6326(17)60060-8

[26] Zhu Y, Chen X, Yang H. Sliding wear of selective laser melting processed Ti_6Al_4V under boundary lubrication conditions. Wear. 2016;**36**(8, 369):485-495. DOI: 10.1016/j.wear.2016.09.020

[27] Patila AS, Hiwarkara VD, Verma PK, Khatirkarc RK. Effect of TiB_2 addition on the microstructure and wear resistance of Ti-6Al-4V alloy fabricated through direct metal laser sintering (DMLS). Journal of Alloys and Compounds. 2019;**777**:165-173. DOI: 10.1016/j.jallcom.2018.10.308

[28] Liverani E, Toschi S, Ceschini Km Fortunato A. Effect of selective laser melting (SLM) process parameters on microstructure and mechanical properties of 316L austenitic stainless steel. Journal of Materials Processing Technology. 2017;**249**:255-263. DOI: 10.1016/j.jmatprotec.2017.05.042

[29] Zhu Y, Zou J, Chen X. Yang H; Tribology of selective laser melting processed parts: Stainless steel 316L under lubricated conditions. Wear. 2016;**350-351**:46-55. DOI: 10.1016/j.wear.2016.01.004

[30] Li H, Ramezani M, Li M, Ma C, Wang J. Tribological performance of selective laser melted 316L stainless steel. Tribology International. 2018;**128**:121-129. DOI: 10.1016/j.triboint.2018.07.021

[31] AlMangour B, Grzesiak D, Yang JM. Rapid fabrication of bulk-form TiB_2/316L stainless steel nanocomposites with novel reinforcement architecture and improved performance by selective laser melting. Journal of Alloys and Compounds. 2016;**680**:480-493. DOI: 10.1016/j.jallcom.2016.04.156

[32] Lorusso M, Manfredi D, Calignano F, Ambrosio EP, Trevisan F, Pakkanen J, et al. Wear behavior of aluminum matrix composites by DMLS reinforced with micro- and nano-TiB_2; In Proceeding of the European Congress and Exhibition on Advanced Materials and Process (EUROMAT2015); Warsaw, Poland. 2015

Chapter 7

Tribological Characteristics of Smart Materials (Magneto-Rheological Fluids and Elastomers) and Their Applications

Peng Zhang, Chenglong Lian, Kwang-Hee Lee and Chul-Hee Lee

Abstract

Magneto-rheological fluids (MRFs) and magneto-rheological elastomers (MREs), as smart materials, have been widely studied in various engineering fields to address vibration issues because the mechanical properties are controllable under the strength of a magnetic field. Their tribological characteristics are also important to be evaluated, as applications using MRFs and MREs contain various contact interfaces under reciprocating and rotating working conditions. The performance and durability of these materials are related to their tribological characteristics. Therefore, various working conditions and environmental conditions are taken into consideration, and their tribological characteristics are experimentally examined. In addition, applications using MRFs and MREs are introduced, and the tribological performances of these materials are evaluated.

Keywords: tribology, magneto-rheological fluid, magneto-rheological elastomer, smart material, friction control

1. Introduction

Magneto-rheological fluids (MRFs) and magneto-rheological elastomers (MREs) have been extensively studied to solve vibration problems in various engineering fields. They have one or more attributes that can be significantly changed in a controlled way by external stimuli (magnetic fields). It is essential to evaluate the tribological properties as they relate to the performance of smart material-based applications.

MRF consists of base fluid with magnetic particles forming chain shape along the magnetic field direction. Because of its fast response speed, it has the potential to be applied to various industrial sectors such as automotive, aviation, construction, etc., [1]. Although research on MRF-based applications with control method is being conducted, tribological characteristics of MRF remain in early stage.

For example, tribological properties of magnetic particles in MRF are examined by Bullough [2]. It is noted that the parameters such as particle concentration,

IntechOpen

surface condition, sliding speed, and contact pressure should be considered to understand sliding contact mechanism. Lubrication performance of MRF is evaluated using tribological tester [3].

The performance of MRF is based on the yield stress, which can be changed by the strength of a magnetic field [4]. The applied magnetic field can change the intensity of the particles arranged along the magnetic field direction. The particles can affect the friction and wear at the contact surfaces. Furthermore, the environmental conditions can affect the properties of fluid, and the changed liquid properties can also affect friction and wear. Fridrich studied that the heated surface leads to the high specific wear rate with lower friction coefficient [3]. Most research focuses on the tribological characteristics of MRF without the magnetic field [2, 5].

Since there is a disadvantage that the environmental pollution problem due to leakage can occur, it is essential to apply the technology to prevent leakage of the fluid. Performance degradation due to particle deposition in the fluid is also a major problem. To solve these drawbacks, MRE has been proposed. It consists of base material (generally polymer) and magnetic particles. As with MRF, mechanical properties can be varied depending on the presence or absence of a magnetic field [6]. Studies on the friction and wear characteristics of ordinary rubber have been actively conducted, but the research on MRE is still insufficient [7].

MREs, as smart elastomers, are also investigated for use in various types of equipment [8]. Specifically, MREs are also used in various external environments at different temperatures, relative humidity conditions [9], and vibration conditions [10]. However, there is a paucity of studies examining the friction and wear properties of MREs. The friction coefficient of an MRE can be controlled using external magnetic fields by changing the hardness of the MRE. Therefore, this property can be applied in some hardness-controllable devices.

MRFs and MREs have been widely studied and applied in various mechanical devices. Because of their controllable mechanical properties under the magnetic field, their tribological characteristics are important for evaluation. However, in this chapter, the tribological characteristics of MRFs and MREs will be explained, including friction and wear properties of MRFs and MREs under different environmental conditions such as temperature, humidity, and vibration conditions.

2. Tribological characteristics of magneto-rheological fluids (MRFs) and their applications

2.1 Friction and wear properties of MRF under different applied loads and speeds

The potential applications using MRF are operated under various loads and speed conditions. It is necessary to estimate the performance of applications that take into account friction and wear characteristics. As most contacts are occurred in linear motion, friction and wear characteristics are examined under different working conditions (oscillating frequency, load, and magnetic field). The detailed mechanism of the tester is shown in **Figure 1**. The magnetic field is applied to the MRF, and its direction is perpendicular to the moving direction. Cartridge heaters under the specimen can provide the heat on the contact surface. The oscillating frequency and the applied load are fixed to 0.5/1.0 Hz and 1/5 N, respectively. The results in **Figure 2** show the friction coefficient changes concerning the working cycles. The lower values are observed under higher load condition with no magnetic field. The tendency of the results with different oscillating frequencies is almost the same.

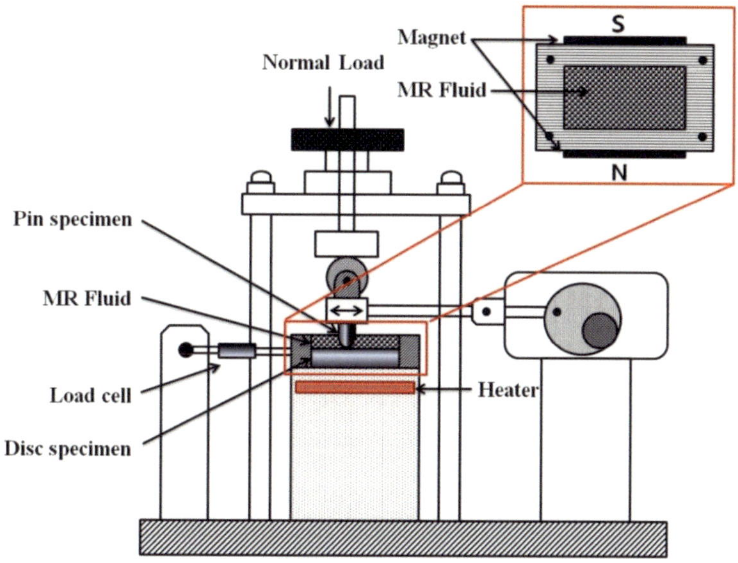

Figure 1.
Schematic of reciprocating friction and wear tester (R&B 108-RF).

The microscopic images of the specimen before and after the tests are compared as shown in **Figure 3**. The machining marks are only observed on the surface before the tests, but some wear marks such as ridges are shown on the surface after the tests. The distance between wear marks is very close, and a smoother surface is observed when a magnetic field is applied. The particles in MRF form a chain shape along the direction of a magnetic field. The formed structure works as resistance resulting in small motion at contact interfaces. However, the main reason for surface wear is free-moving particles.

2.2 Friction and wear properties of different types of MRFs

Since the mechanical properties of MRF are controllable, various types of MRF (122EG, 132DG, 140CG) are studied to improve the MR effects. The weight percentage of particles in the fluid is one of the key factors. The particles in the fluid affect friction and wear properties at contact interfaces. Therefore, various MRF types are taken into consideration to estimate the tribological characteristics under different working conditions such as the strength of the magnetic field, load, and oscillating frequency. The magnetic field strength is changed to 9 mT from 3 mT with a 3 mT step. Moreover, oscillating frequency and the applied load are fixed to 1 Hz and 10 N, respectively.

The results of the friction coefficient change with different magnetic field strengths are shown in **Figure 4**. The friction coefficient tends to increase as the magnetic field strength increases. The particles are constrained in the chain shape along the direction of the applied magnetic field. Some may have an abrasive effect on the surface during the friction process. Most of them show the resistance on the surface due to the gathered particle under the presence of a magnetic field. The particles are turned into the additional tribological pair and increase the friction coefficient. Moreover, the interactions among the particles under the magnetic field continuously occur during the movement, and it causes energy consumption, resulting in friction coefficient increase. Such an interaction force tends to increase under higher magnetic field strength.

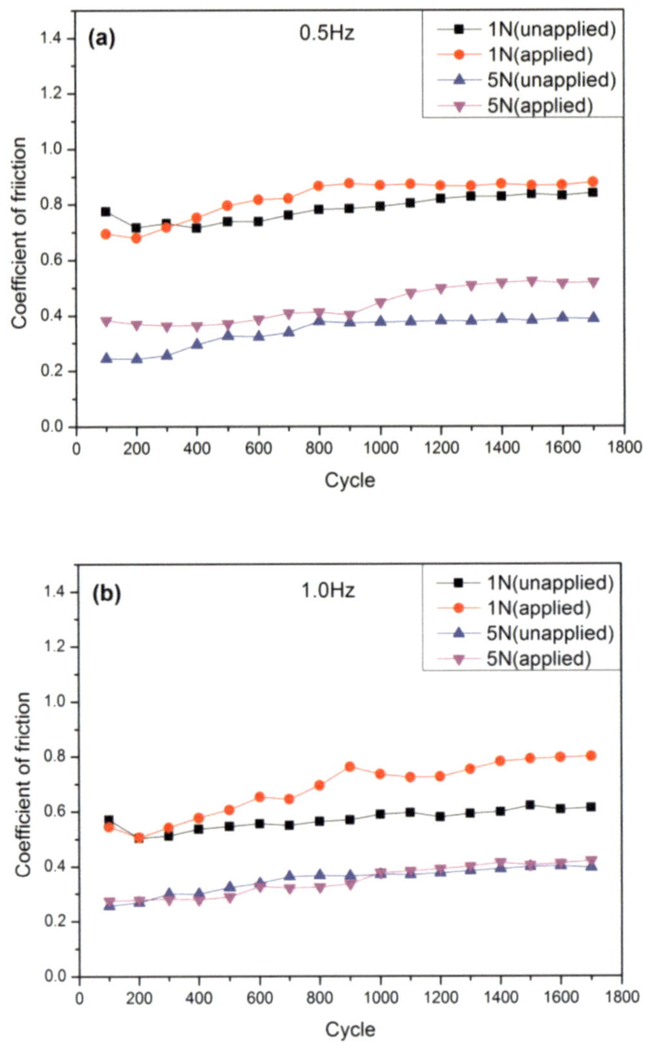

Figure 2.
Friction coefficient change of MRF under different loads and speed conditions (aluminum).
(a) 0.5 Hz/unapplied and applied magnetic field and (b) 1.0 Hz/unapplied and applied magnetic field.

2.3 Friction and wear properties of MRF with the coatings

Since the types of MR fluid have influence on the tribological properties, the demand for the improved friction and wear performance in contact interfaces is increasing. The surface coating is the solution that can enhance the friction and wear performance while maintaining the controllable properties of MR fluid in various industrial applications. Various coating materials are taken into consideration to evaluate the friction and wear properties of MRF. Most common coating materials widely used in industrial area are PTFE and DLC. Therefore, two different coating materials and pairs are considered with the fixed working conditions. The oscillating frequency, applied load, the strength of a magnetic field, and temperature are 1 Hz, 5 N, 5 mT, and 25°C, respectively.

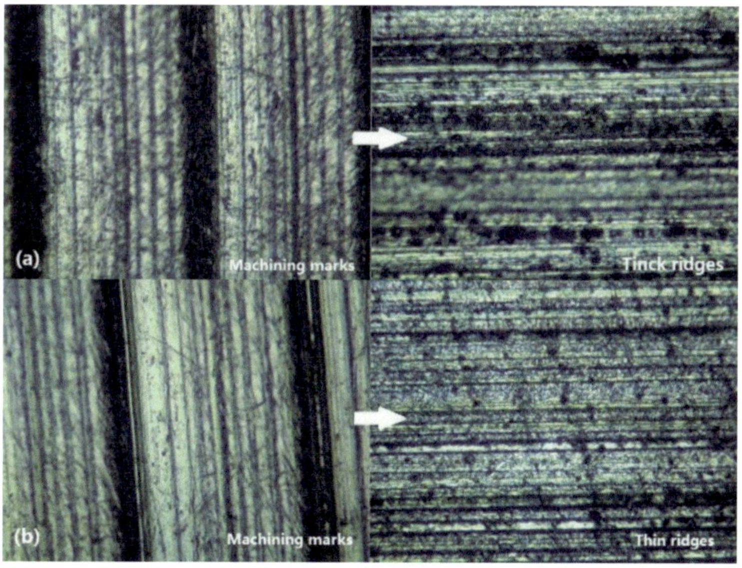

Figure 3.
Microscopic surface images before and after wear (5 N/1 Hz). (a) magnetic field on and (b) magnetic field off.

Figure 5 shows the friction coefficient changes with different coating materials and pairs. The material of the specimen is aluminum (pin and plate), and they are coated with DLC and PTFE. The friction coefficient tends to increase with no coatings. However, it shows the lower values with coatings. The stable results are shown for the coating surface. PTFE-coated surfaces show improved friction performance compared to DLC and no coating. The best friction performance is observed for PTFE coating pairs.

DLC coating is widely used for industrial applications due to the better mechanical properties than PTFE. Thus, it is suggested that the DLC coatings are more appropriate to most of the applications. The temperature is another factor that affects friction performance as the properties of fluid change at higher temperature conditions. The specimen with DLC coating under different temperature conditions is tested. The results of the friction coefficient change with temperature up to 130°C are shown in **Figure 6**, and it tends to decrease as the cycles increase. The degraded viscosity of MRF due to the temperature increase gives more freedom to the particles in the fluid resulting in the improved friction performance [9].

2.4 Friction and wear behavior of MRF applied to pin-bushing system

Bushings are widely used in many mechanical applications working as bearings. Therefore, it is important to either maintain or enhance the tribological properties of bushings. MRF can be applied as a lubricant to bushings to enhance the load-carrying capacity. The friction performance of bushing is evaluated using specially designed tester shown in **Figure 7**. This tester can adjust the oscillating frequency, angle, and load. Friction performance of MRF is tested with a step load up to 5kN from 1kN with and without a magnetic field. The results are compared to the one with conventional grease. In **Figure 8**, the friction coefficient tends to decrease when the load is 3kN and becomes stable. The highest friction coefficient is shown when a magnetic field is activated during the test. It is assumed that the particles in the fluid work as a solid lubricant in the shape of the chain in the direction of a magnetic field. The force among the particles leads to resistance during the operation. The friction performance at the contact interface can be controlled with a magnetic field. The load-carrying capacity also can be changed with MRF.

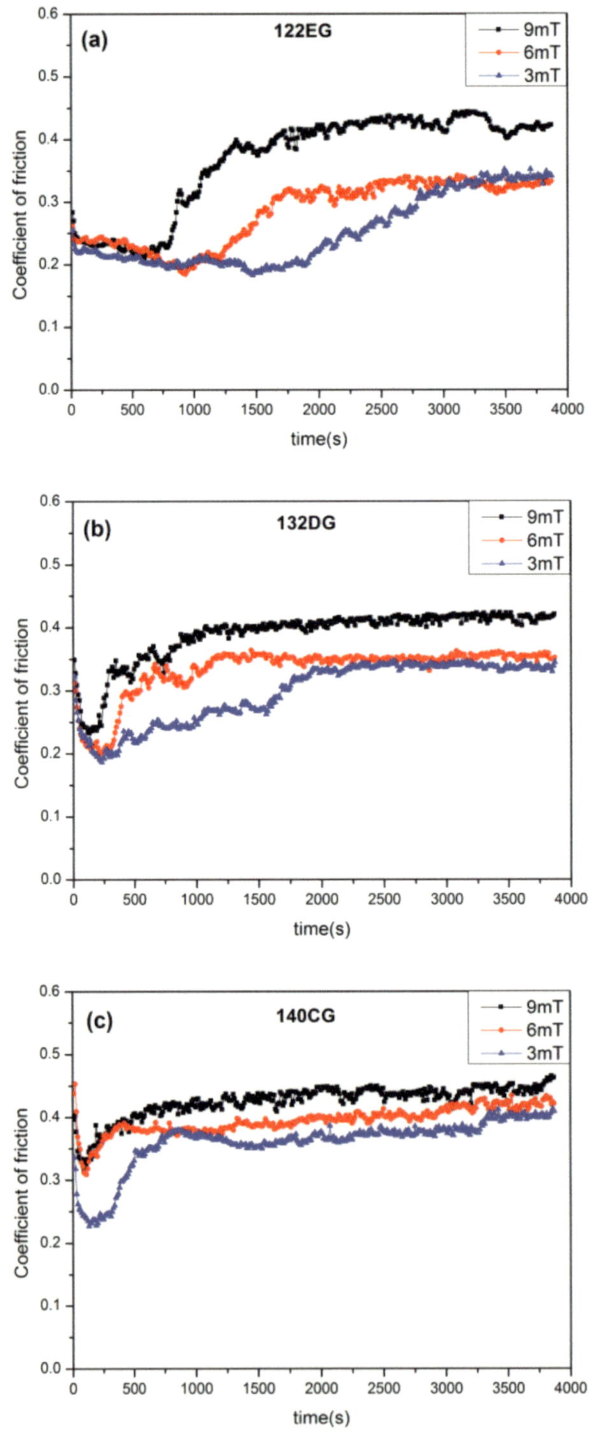

Figure 4.
Friction coefficient changes with respect to types of MRF and strength of magnetic field. (a) MRF, 122 EG; (b) MRF, 132 DG; and (c) MRF, 140 CG.

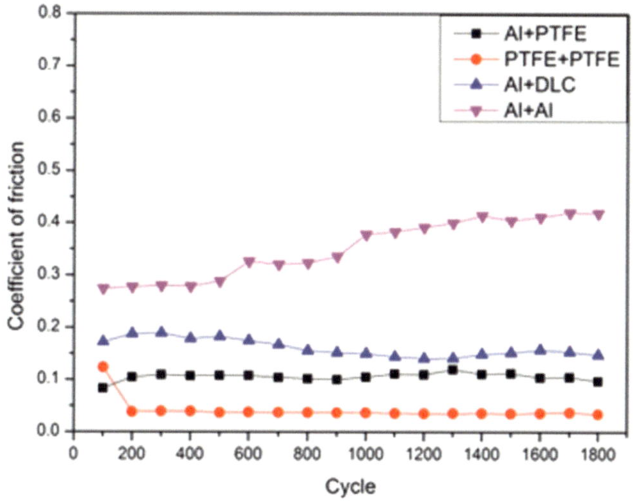

Figure 5.
Friction coefficient changes of MRF with various coating surfaces (5 N/1 Hz).

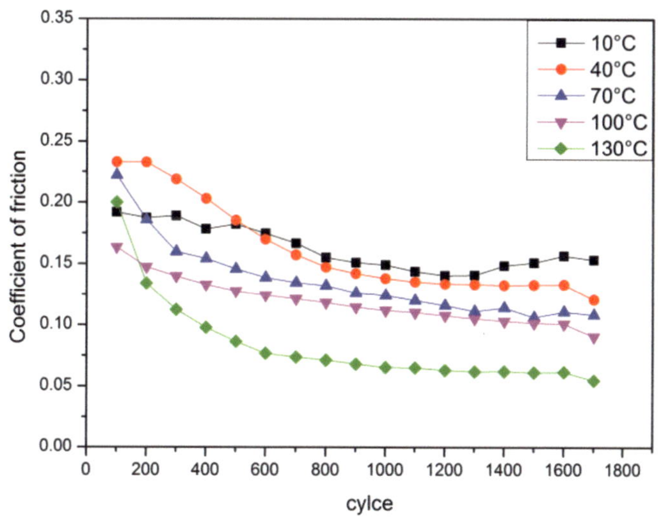

Figure 6.
Friction coefficient changes of MRF with various temperature conditions on DLC-coated surface (5 N/1 Hz).

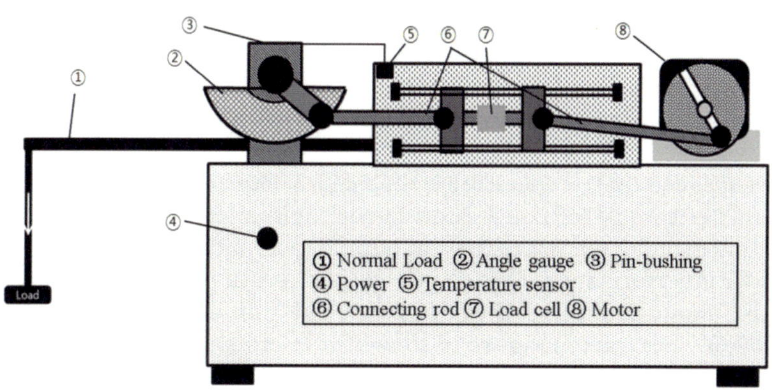

Figure 7.
Schematic diagram of pin-bushing friction and wear tester.

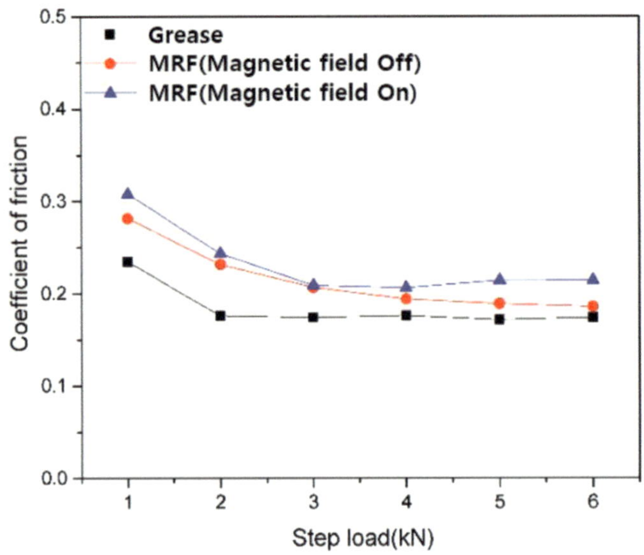

Figure 8.
Friction coefficient chance under step-load condition.

3. Tribological characteristics of magneto-rheological elastomer (MRE) and its applications

3.1 Friction and wear properties of MREs under different temperature conditions

MREs are new types of a smart material that can change mechanical properties under a magnetic field. Like other smart materials, MREs should maintain its durability under severe environmental conditions. As MRF shows different friction characteristics under different temperature conditions, MRE should be tested under the same conditions. The friction and wear performance of MRE surfaces are evaluated on the temperature up to 100°C from room temperature. **Figure 9** shows the results of the friction coefficient and wear depth changes concerning the temperature increases. Friction coefficient tends to increase as the temperature increases, and higher values are observed with no magnetic field. The presence of a magnetic field increases the hardness. The difference in wear depth is larger than the friction coefficient. The friction and wear performance of MRE is affected by the temperature condition. It seems that lower resistance is observed at high-temperature conditions. In the case of fixed load and velocity conditions, the temperature is the major factor of wear at contact interfaces.

3.2 Friction and wear properties of MRE under different humidity conditions

It is mentioned that temperature is the major factor related to the friction and wear characteristics of MRE. It is considered that humidity is another factor. The same tests are conducted with different humidity conditions to estimate friction and wear performance. The test conditions were implemented by adjusting the relative humidity through the humidifier. The humidity conditions are 40, 60, and 80%. The result in **Figure 10** shows the friction coefficient and wear depth change according to the humidity variation. Additionally, the results are compared with or without a magnetic field. Higher values of friction coefficient appear when a magnetic field is applied, and they tend to decrease as the

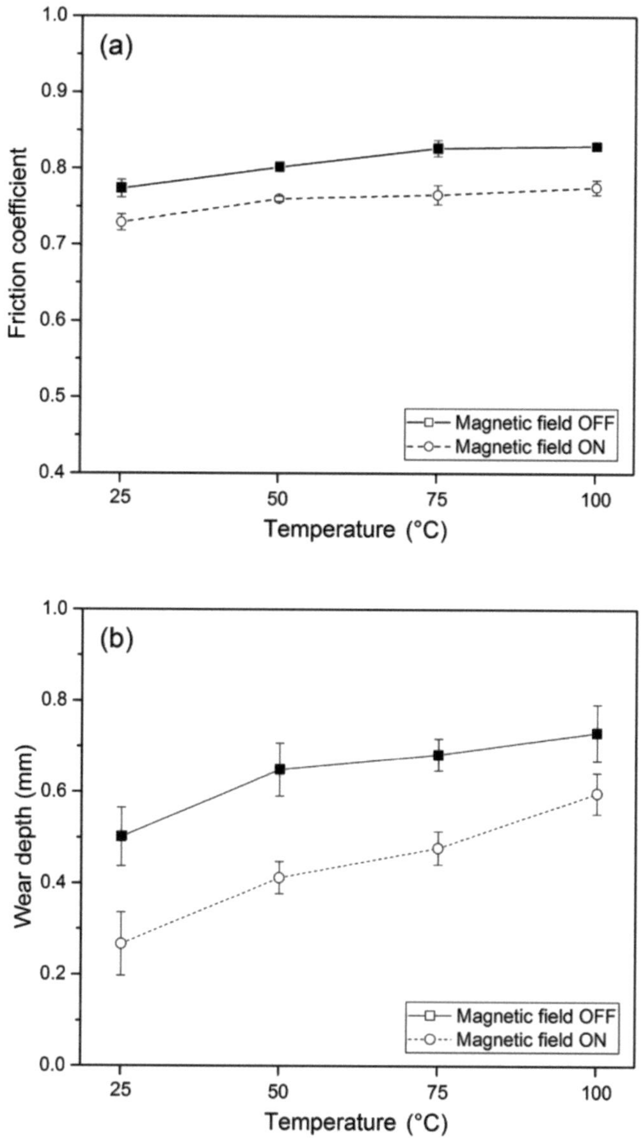

Figure 9.
Results of (a) average friction coefficient and (b) wear depth for MRE under different temperature conditions.

humidity increases. The friction coefficient difference remains almost constant under magnetic field conditions. The applied magnetic field changes the hardness. The reason for the lower friction coefficient under high humidity condition is that the contact surface is affected by the local hydrodynamic effect resulting in a lubricating effect. Also, the water is absorbed in MRE, and its shear strength seems to be decreased.

The wear depth is also reduced as the humidity increases. It is assumed that the shear strength of the contact surface can be reduced by the hydrodynamic effect. This leads to lower energy consumption during frictional movement. The results of wear depth show that severe wear does not appear when the humidity reaches to 80% regardless of a magnetic field. Furthermore, it does not show the significant change of wear depth when the humidity is 60%, which is assumed to be a saturation point.

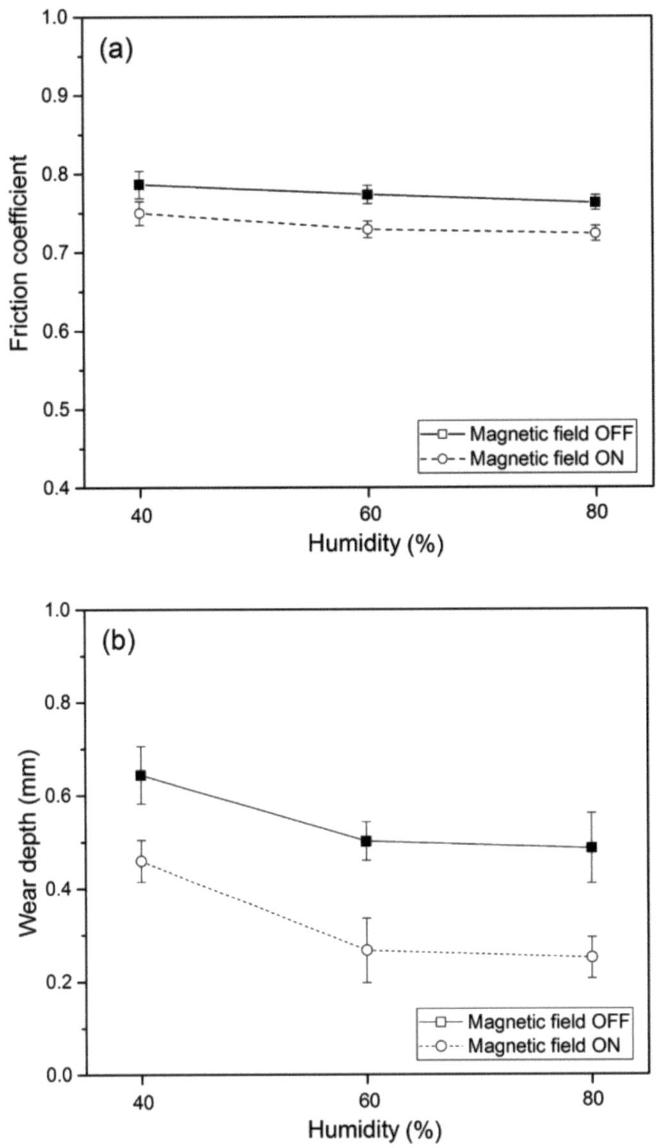

Figure 10.
Results for average friction coefficient (a) and wear depth (b) of MRE at different relative humidity conditions.

3.3 Friction and wear properties of MRE under different vibration conditions

Most industrial applications have vibration issues during the operation. Such vibration is correlated to friction and wear. Previously, friction and wear properties are affected by various environmental conditions. Additionally, it is necessary to estimate the tribological performance of MRE under vibration condition. The tests are conducted under various frequencies and amplitude conditions. The results are shown in **Figure 11**. The average friction coefficients are obtained, which are compared each other, to analyze how the vibration frequency and amplitude affect.

The friction coefficient tends to increase as the frequency increases except at the resonant frequency. Elastomers show different tribological properties,

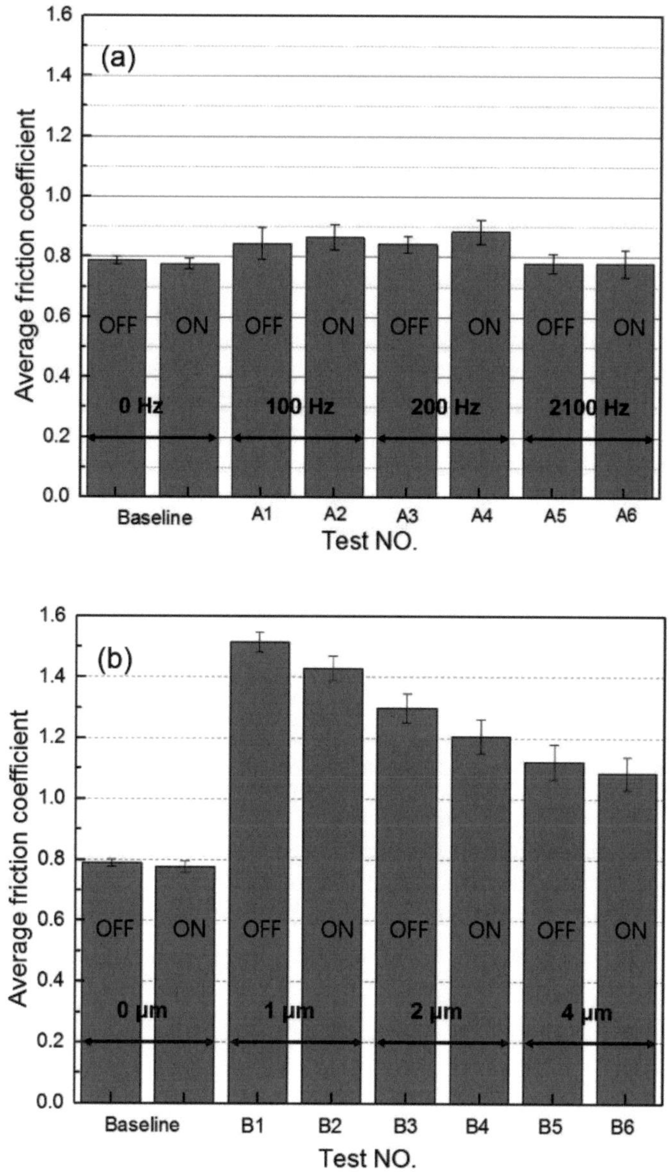

Figure 11.
Average friction coefficient results of MRE under (a) different vibration frequencies and (b) vibration amplitudes. A1, A3, and A5: 100, 200, and 2100 Hz without a magnetic field. A2, A4, and A6: 100, 200, and 2100 Hz with a magnetic field. B1, B3, B5: 1, 2, 4 μm without a magnetic field. B2, B4, and B6: 1, 2, and 4 μm with a magnetic field.

unlike metal material in which friction coefficient reduces under vibration conditions [10]. The surfaces of elastomer have less resistance as temperature increases, reducing the friction coefficient. When high-frequency oscillations occur, vibration energy is converted into heat energy. The temperature of MRE is increased by 2°C when a vibration is applied. It is assumed that the generated heat on the surface leaded higher friction coefficient under vibration conditions. Also, the increased contact time under high frequency causes a higher friction coefficient. The lower values are observed at the conditions of 100 and 200 Hz. After 200 Hz, the vertical load at the contact surface is reduced by the resonance. The lower friction coefficient can be obtained due to the reduced load. The vibration amplitude is taken into consideration to evaluate the tribological properties of MRE. The previous results show that the friction coefficient clearly decreases when a magnetic field is applied, which is related to the hardness change. The small deformation on the contact surface appeared when a magnetic field is applied, resulting in small deformation; when a vibration amplitude increases, the friction coefficient decreases. It is assumed that the separation of the contact pair appears when the amplitude is high.

3.4 Rolling friction characteristics of MRE

Previously, the friction coefficient under reciprocating motion can be controlled by the strength of a magnetic field. Rolling friction is another factor to be considered in engineering applications. As the motion is different, tribological characteristics of MRE under rolling motion should be taken into consideration. The tests are carried out under a fixed velocity and load conditions.

The micro slip at the contact interfaces, adhesion, and plastic deformation are the key factors of rolling friction of elastomer. The schematic of the rolling friction of elastomer is shown in **Figure 12**. In the analytical model [11], the rolling friction coefficient can be obtained as follows:

$$\mu_R = \frac{M_R}{WR} = \alpha\frac{2a}{3\pi R} = \frac{4\alpha}{3\pi}\left\{\frac{W}{\pi R E^*}\right\}^{\frac{1}{2}} \quad (1)$$

According to Eq. (1), the rolling friction coefficient can be reduced by increasing the modulus. It can be increased under a magnetic field. Thus, the theoretical values of rolling friction coefficient can be obtained. The theoretical and experimental values are compared in **Figure 13**. The results show that the values tend to decrease as the magnetic field strength increases. The tendency of experimental results is similar to the one in theory. The difference between experiment and theory is caused by micro slip and adhesion at contact interfaces.

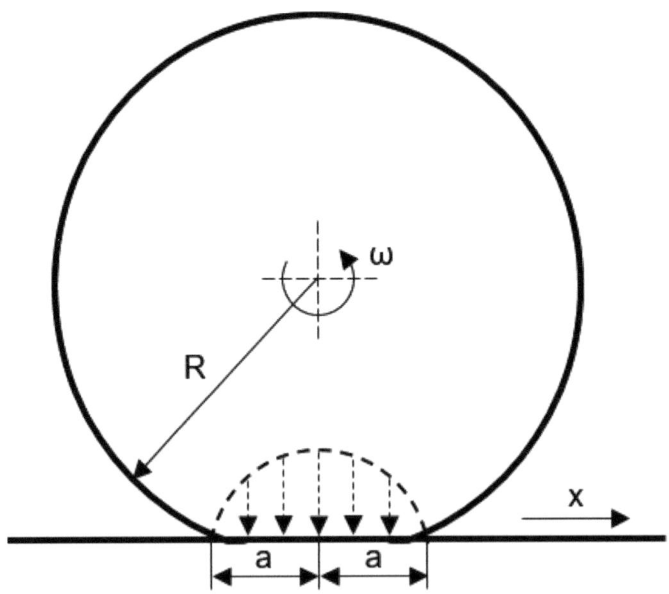

Figure 12.
Analytical model of rolling friction of an elastomer.

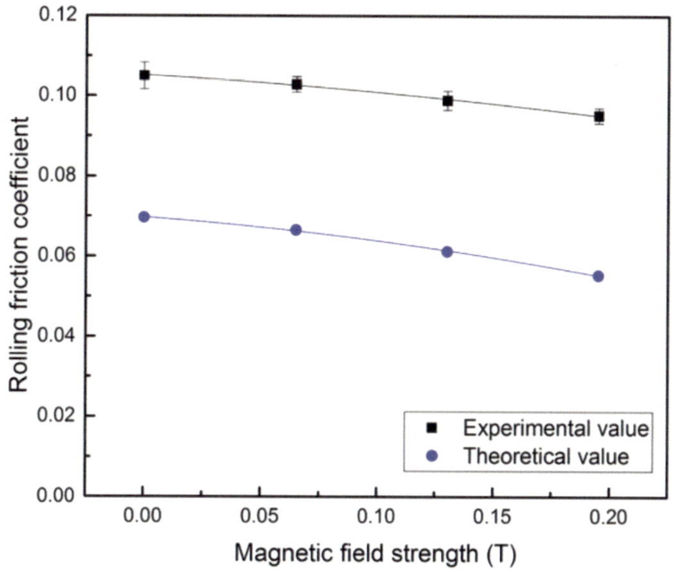

Figure 13.
Comparison of rolling friction coefficient of MRE with respect to magnetic field strength from experiment and theoretical analysis.

Author details

Peng Zhang[1], Chenglong Lian[2], Kwang-Hee Lee[3] and Chul-Hee Lee[3]*

1 Nanjing Research Institute for Agricultural Mechanization Ministry of Agriculture, Nanjing, PR China

2 College of Forestry, Shandong Agricultural University, Taian, PR China

3 Department of Mechanical Engineering, Inha University, Incheon, South Korea

*Address all correspondence to: chulhee@inha.ac.kr

References

[1] Bullough W, Wong P, Fen C, Leung W. Fundamental boundary tribology: ESF. Journal of Intelligent Material Systems and Structures. 2003;**14**:71-78. DOI: 10.1177/104538903030330

[2] Wong P, Bullough W, Feng C, Lingard S. Tribological performance of a magneto-rheological suspension. Wear. 2001;**247**:33-40. DOI: 10.1016/S0043-1648(00)00507-X

[3] Song W, Cai Q, Choi S, Lee C. A study of finishing process of magneto-rheological fluid on steel surface. Civil Engineering and Building Materials. 2011;**1**:17-24

[4] Zhang P, Lee K, Lee C. Reciprocating friction characteristics of magneto-rheological fluid for aluminium under magnetic field. Transactions of Nonferrous Metals Society of China. 2014;**24**:171-176. DOI: 10.1016/S1003-6326(14)63044-2

[5] Zhang P, Lee K, Lee C. Friction behavior of magnetorheological fluids with different material types and magnetic field strength. Chinese Journal of Mechanical Engineering. 2016;**29**:84-90. DOI: 10.3901/CJME.2015.1126.139

[6] Carlson J, Jolly M. MR fluid, foam and elastomer devices. Mechatronics. 2000;**10**:555-569. DOI: 10.1016/S0957-4158(99)00064-1

[7] Lancaster J. A review of the influence of environmental humidity and water on friction. lubrication and wear. Tribology International. 1990;**23**:371-389. DOI: 10.1016/0301-679X(90)90053-R

[8] Persson B. Rubber friction and tire dynamics. Journal of Physics: Condensed Matter. 2010;**23**:015003. DOI: 10.1088/0953-8984/23/1/015003

[9] Odfalk M, Vingsbo O. Influence of normal force and frequency in fretting. Tribology Transactions. 1990;**33**:604-610. DOI: 10.1080/10402009008981995

[10] Wen S, Huang P. Principles of Tribology. Beijing: John Wiley & Sons; 2012. DOI: 10.1002/9781118062913

[11] Williams J. Engineering Tribology. Oxford city: Cambridge University Press; 2005. DOI: 10.1017/CBO9780511805905

Section 2

Lubrication

Chapter 8

A Review of Surface Treatments for Sliding Bearings Used at Different Temperature

Jun Cao, Liang Qin, Aibing Yu, Haibo Huang, Guoping Li, Zhongwei Yin and Huiyu Zhou

Abstract

The boundary lubrication and dry friction of plain bearings at different work temperature are unable to be avoided under the start and stop condition. The poor lubrication is one reason of bearing broken. In order to improve the tribological properties and select the best treatment for different bearings used at different temperature, the studies of different treatment technologies are reviewed in this paper. The review shows that the shortages of bonding fiber woven materials, inlaying solid lubricating materials, electro plating and magnetron sputtering are poor temperature resistance, low load capacity, environment pollution and low production efficiencies respectively. Based on the analyses and summaries, the liquid dope spraying and thermal powder spraying are suggested to deposit coating on the surface of bearing which working temperature is lower than 200 and above 800°C respectively. However, the technology processes, the mechanisms of spraying and self-lubrication materials should be studied further and deeply.

Keywords: plain bearing, surface treatment, wide range temperature, lubrication materials, technology process

1. Introduction

1.1 Background of the research review

Sliding bearings are widely used as the basic components in marine power, aerospace, water conservancy and hydropower industries. As shown in **Figure 1**, the sliding bearings are divided into integral sliding bearings and split sliding bearings according to their structures. They have some features, such as low noise, stable work operation, compact structure and heavy load bearing capacity [1]. With the advancement of science and technology, the automotive, marine, electric power transportation and some other industries are developing into 'serious conditions' such as high speed and high load, making the sliding bearing under a wide temperature environment. For example, the work temperature of engine rises from room temperature to about 140°C during different operating conditions, such as starting, accelerating, constant speed, deceleration and shutdown. From the research data, the temperature even rises up to 200°C at the time of bearing broken [2]. The work temperature of some bearings is high, for example, the working temperature of the socket bearing of missile launching mechanism exceeds 800°C. Under high temperature conditions, grease and lubricating oil fail, and unlubricated bearings will quickly be broken under the action of high friction coefficient and wear. Ordinary sliding bearings are normally lubricated due to lubricating oil with the hydrodynamic lubricating oil film. Under normal conditions, sliding bearings can generally satisfy the requirements of long-term service. However, the reasons of ablation and friction damage of the bearings 12.5% are due to poor lubrication the investigation.

As shown in **Figure 2**, the sliding bearing damage quickly due to the lacking of self-lubricating properties of bearing, and the mixed lubrication and even dry contact friction at the start-stop stage of engine [3]. Therefore, researching and improving the friction mechanical properties of the bearing under dry friction conditions is one of the key technologies for improving the bearing life of at start-stop stage. In the case of large dust, high temperature and no lubricating oil or grease, the life of the sliding bearing will be drastically reduced. For example, the working environment temperature of rolling steel rolling mill is about 200°C, and the maximum service life of the bearing is no longer than 3 months. In order to avoid the bearing damage and improve the service life of the sliding bearing at the dry friction stage, it requires the sliding bearing should have self-lubricating property to reduce the high torque requirement and tribology at the stages of start-stop under different temperature conditions [4].

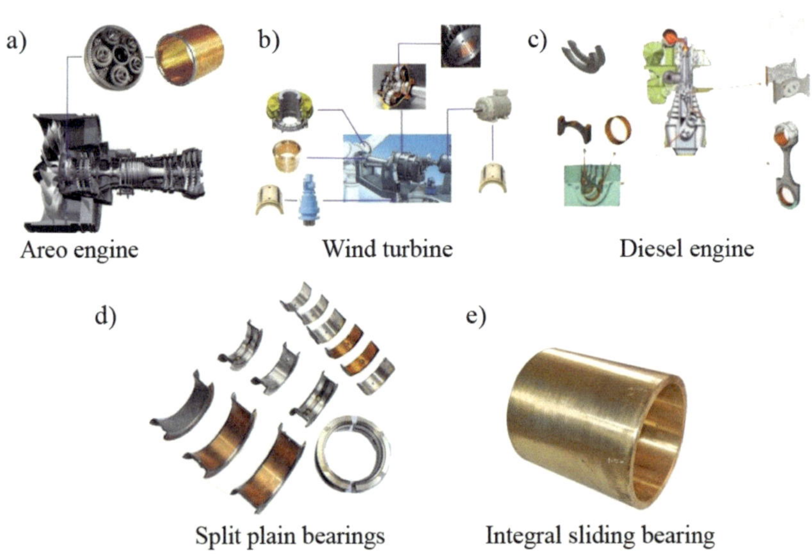

Figure 1.
The applications of plain bearings and their structures.

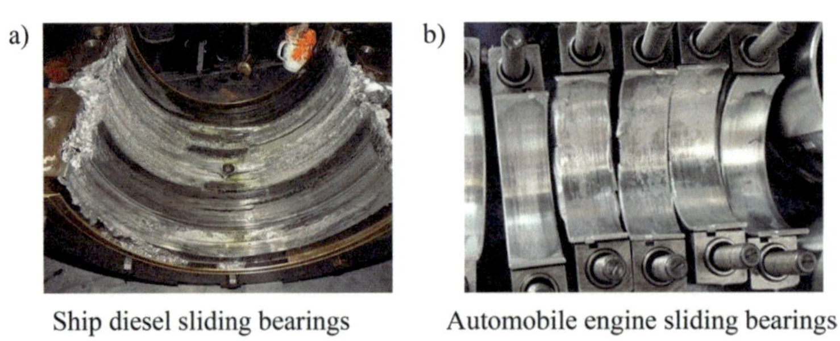

Figure 2.
Friction damage of sliding bearings.

The self-lubricating treatment technology of sliding bearings in China is still backward. Bearings with self-lubricating properties at different temperature conditions are still very rare. Compared with the advanced bearings of European and North American, the sliding bearings of China have short life, low carrying capacity, poor self-lubricating performance and overcapacity. High-performance sliding bearings such as, self-lubricating, high-load, and long-life are relied on import. The lack of high-performance sliding bearings restricts the development of China's basic manufacturing industry, especially the military industry. Therefore, the self-lubricating performance of the sliding bearing during in wide temperature range should be improved, and the different lubricating methods for different bearings at different temperature should be selected.

The development, manufacturing and processing capabilities of sliding bearing materials of China should be improved. At present, the main production technologies sliding bearings are centrifugal casting and alloy powders metallurgical. However, most bearings have no surface lubrication and will be quick broken. For example, more than 90% of the bearing bushes prepared by powder metallurgy in the automotive industry are not subjected to surface self-lubricating treatment such as electroplating and magnetron sputtering. Powder metallurgy technology of sliding bearing preparation has high production efficiency and low cost. The sliding bearing prepared by metallurgical process has high porosity without centrifugal force. In addition, the bearing alloys such as copper, aluminum and tin are easily oxidized at high temperature during the metallurgy preparation of sliding bearing, and metallographic organization is not uniform [5]. Centrifugal casting is another common production process for sliding bearings. The integral and thick-walled plain bearings are usually produced by centrifugal casting. However, most of the sliding bearings produced by centrifugal casting process are not subjected to inert gas protection or vacuum environmental protection, therefore, the bearing alloys are oxidized and the alloy grains size are not uniform [6].

With high dry friction coefficient and low load bearing capacity of bearings, the common methods are the surface treatments which improve the bearing life effectively. The electroplating, magnetron sputtering, self-lubricating liner antifriction, and inlaying self-lubricating materials are the main surface treatments for bearings. The lubrication and mechanical properties of the sliding bearing surface are changed by one or several electro-plating alloy layers. However, the plating solution is highly polluted, and electroplating technology of sliding bearings is seeking alternative process technology [7].

The magnetron sputtering is one of the most advanced technologies for the preparation of sliding bearings. Compared with electroplated bearings, magnetron sputtering bearings have better bonding strength and surface lubricity. However, magnetron sputtering requires a process such as pumping, vacuuming, and sputtering to form a uniform film. The magnetron process needs long production time and high production cost, and its target materials utilization rate are lower than 40%, which cannot satisfy the requirements of large-scale production. What is more, due to the constraints such as size and structure, just small bearings are able to be prepared by magnetron sputtering [8]. At present, it is an urgent problem to find a mass production of sliding bearing to satisfy the requirements of self-lubrication at wide temperature range, and having good lubricity under special working conditions such as start-stop, lean lubrication or even dry friction. The materials of electroplated plain bearing are mainly babbitt alloy, ternary and quaternary indium alloys which friction coefficient is large under dry friction conditions, and it does not have wide temperature range self-lubricating performance. The Kevlar aramid fiber modified with nano-solid lubricants, and pasted on the surface of the sliding bearing that is the liner anti-friction technology. The liner and pasted glue cannot be used at high temperature conditions, and it does not have self-lubricating properties at wide temperature. The sliding bearing inlaid solid self-lubricating materials such as graphite, MoS_2, WS_2 and so on, and they are punched on the working surface of the sliding bearing. With self-lubricating materials, the self-lubricating materials are crushed to form a self-lubricating film to reduce the friction coefficient of the bearing. However, single-phase self-lubricating material such as graphite and MoS_2 cannot satisfy the requirements of wide temperature range self-lubrication, and the inlaid holes will reduce the mechanical strength and load capacity of the bearing. The temperature environment of the joint bearing, machine tool and electric equipment sliding bearings is room temperature environment; the working temperature of the sliding bearing for hot-burning furnace, gas pump and rolling steel rolling roller is about 200°C; the working temperature of the bearing socket for aviation is above 800°C, and they require self-lubricating materials at different temperature environments. Studies have shown that the use of coating lubrication technology improves the friction and wear, impact resistance, high temperature and longevity of the sliding shaft without changing the bearing matrix structure and composition [9]. Therefore, bearing surface coating technologies are one of the most critical and feasible methods to improve the overall technology of the domestic sliding bearing industry.

1.2 Significance of the research review

Due to the higher requirements of high-speed, high-load and high-temperature, the lubrication of sliding bearings under different temperature conditions and different load conditions will directly determine the working state and service life of the bearings and the whole machine. Many scholars have studied the self-lubricating methods of sliding bearings such as electroplating, magnetron sputtering, inlaying solid lubricants, and adhesive self-lubricating liner. However, these traditional sliding bearing self-lubricating methods have some defects. With the improvement of environmental protection requirements, the sliding bearings prepared under large-scale and high-volume production conditions have excellent self-lubricating properties under different temperature conditions, which are the keys of the current research. One of the current advanced treatment technologies for self-lubricating sliding bearings is liquid coating technology, but the theoretical calculation of bearing spraying is litter. The phenomenon of bearing sag and leveling has not been studied deeply. The optimum thickness of coating, the best surface roughness of the substrate, and coating the optimum curing temperature and optimum cooling temperature of the layer also lack of relevant details and theoretical analysis. The studies of sliding bearings with wide temperature range self-lubricating coating materials under different temperature conditions are lacking.

2. Reviews of self-lubricating sliding bearing

The traditional self-lubricating sliding bearing production and preparation processes are mainly electroplating, magnetron sputtering, inlaying solid lubricants, and adhesive self-lubricating liner and so on. The self-lubricating bearing processes and their performances are summarized as follows.

2.1 Review of self-lubricating sliding bearings at room temperature

In the severe conditions such as large dust, high pollution, high load bearing, lacking of lubricating oil and grease, sliding bearings only rely on their own lubrication to improve work performance. For example, in the automobile manufacturing, cement production and coal mining industries, the inlaying solid lubricants and adhesive self-lubricating liner are the common methods of the joint bearings and bushing sleeves.

2.1.1 Adhesive liner for self-lubricating bearing

The friction coefficient of the bearing is reduced, the wear resistance is improved and the working life is prolonged through the modified liners for sliding bearings. Braided liners are generally composed of Kevlar aramid fiber materials (KEVLAR), polytetrafluoroethylene (PTFE), modified carbon fiber materials, and nano-additives [10]. The structure of sliding bearing with liner is shown in **Figure 3**. In order to obtain a small shear force and a large bonding strength, the fabric liner is bonded on the bearing surface through the adhesive glue. The frictional coefficient is decreased by changing the metal to metal contact to metal to liner contact. Aderikha studied the friction and wear properties of the liner based on PTFE and plasma treated polymer fibers. The results showed that the friction coefficient was 0.15–0.2 under different loads [11]. Li studied the friction and wear properties of nano-materials SiC and WS_2, and the friction coefficient of surface liner under dry friction was about 0.05–0.06 [12]. Fabrics were treated with rare earths CeO_2, $LaCl_3$, La_2O_3 and CeF_3 by Shen, Zhan and some scholars, and their friction, wear properties and bonding properties of the joint bearings with the modified rare earths were studied. The results shown that the bonding strength was higher under the action of rare earths, the film formation is faster, and the coefficient of friction is generally less than 0.1 [13–16]. Liners and adhesive glue as the main component cannot be used at medium and high temperatures, and the bonding liner method is generally applicable to small thick-walled sliding bearings, which has certain limitations for medium and large sliding bearings.

2.1.2 Bearings with inlaid solid self-lubricating materials

The self-lubricating materials are inlayed on the surface of bearing. As shown in **Figure 4**, the solid lubrications will be expanded when sliding bearing subjected to load, and the bulged out lubrications are ground to tiny wear debris. With the sliding movement of bearing, the frictional coefficient is decreased with the formation of lubricating film from the debris. The inlaid materials are generally the solid lubricant materials such as graphite, molybdenum disulfide and PTFE. Wei prepared a new self-lubricating material consisted of PTFE, graphite and glass fiber. The results showed that the friction and wear properties of self-lubricating materials prepared by 40% PTFE + 20% graphite + 20% lead powder + 20% glass fiber were the best [17]. MoS_2/Sb_2O_3 mixed powders were produced to form solid

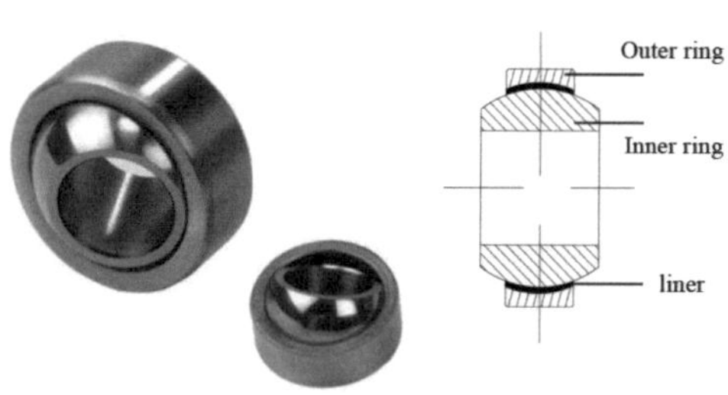

Figure 3.
Join bearing with self-lubricating liner.

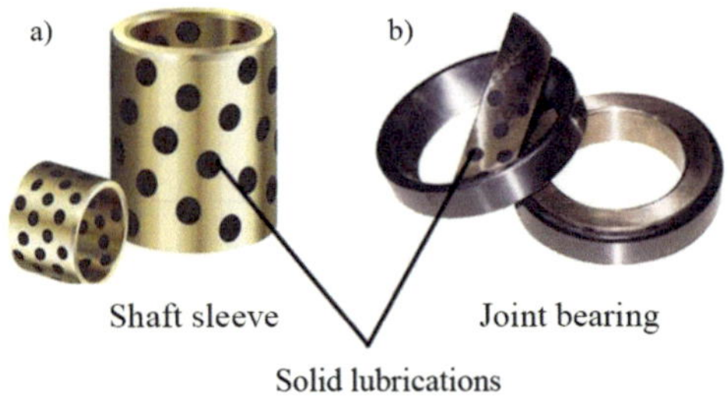

Figure 4.
Plain bearing inlaying with solid lubrications.

lubricant by Zabinski with thermosetting bonding method. The results pointed out that MoS_2 and Sb_2O_3 have synergistic antifriction effect on friction work, and Sb_2O_3 can prevent MoS_2 from oxidizing [18]. Li prepared FeS/copper-tin alloys as the inlaying materials by powder metallurgy. The research showed that the increase of FeS content reduce the friction coefficient. When the FeS content is 10%, the friction coefficient is 0.15 [19]. The research studies of inlaid solid self-lubricating materials showed that the working conditions of the prepared materials were mostly room temperature environment, and the studies of medium and high temperature self-lubricating materials were few. Therefore, the working conditions of sliding bearings embedded with solid self-lubricating materials on the market are mostly room temperature environments. In addition, the inlaid structure will reduce the strength of the bearing, resulting in low bearing capacity.

Sliding bearings prepared by modified liner technology and inlaid solid self-lubricating technology satisfy the self-lubricating requirements of room temperature conditions, but the fiber fabric materials cannot be used at high temperature. However, there are few studies of solid self-lubricating materials that satisfy the self-lubricating under the wide temperature range. Generally, bearings with inlaying solid materials are used at room temperature environment, and the structural strength of the bearing will be reduced by this technology.

2.2 Review of self-lubricating sliding bearing at medium temperature condition

2.2.1 Electroplating

The working temperature of bearing bush, plain bearing of rolling mills near to furnace and sleeves is from 100 to 200°C. At the start and stop stages, the automobile bearing bush is under a boundary lubrication or even a dry friction state because the lubricating oil film is not formed at start stage or broken at stop stage. However, the sliding bearings of rolling steel and the bearings in the heating furnace may cause the lubricating oil and grease to fail due to the high working environment temperature. In order to improve the performance of the bearing and prolong its service life, bearings need to be self-lubricated. The most common self-lubricating treatment methods for plain bearings represented by bearing bushes are electroplating and magnetron sputtering. As shown in **Figure 5**, one or more layers can be prepared on the surface of the bearing to improve bearing lubrication and improve bearing fatigue strength and service life. Wang studied Ni/SiC and Ni/Al_2O_3 electroplating techniques, and ceramics such as SiC and Al_2O_3 were added to the coating material, which improve the wear resistance of the bearing [20]. Li studied the friction and wear behavior of nano-Ni-PTFE composite coating on steel substrate. The results showed that the coating friction coefficient range was 0.05–0.15 under different loads [21]. Zhang prepared a MoS_2 coating containing

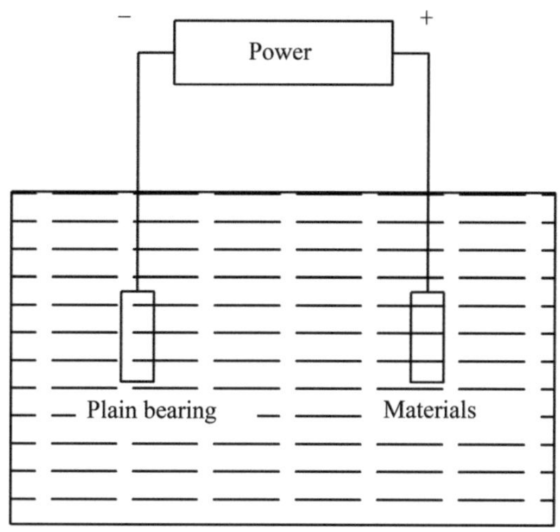

Figure 5.
Technology of electroplating of bearings.

nano-graphite particles by electroplating brush, and tested the friction coefficient was from 0.05 to 0.15 [22]. However, the plating prepared by the electroplating brush is not uniform, and the bonding strength is not as good as that of chemical electrophoresis. Studies had shown that metals such as In, Ni, and W improve the wear resistance of sliding bearing coatings [23, 24]. The addition of rare earth metals such as La, Ta, Nb significantly improved the frictional mechanical properties of the sliding bearing coating [25–27]. The friction coefficient of electroplated copper-tin alloy, aluminum-tin alloy and babbitt alloys under dry friction conditions is generally 0.3–0.6, which needs to be combined with lubricating oil and grease to satisfy the lubrication requirements [28]. Electroplated plain bearings are currently the most widely used preparation methods, but the plating solution is highly polluting and does not satisfy environmental production requirements.

2.2.2 Magnetron sputtering

Unlike electroplating, magnetron sputtering does not cause environmental pollution. The magnetron sputtering process shown in **Figure 6** has a dense film, and the thin metallographic structure in a vacuum environment makes the performance of the sliding bearing superior to that of the electroplated production. Li prepared a Babbitt Cu-Sn-Sb film on a steel substrate by magnetron sputtering. The friction coefficient was from 0.1 to 0.25 after dry friction experiment of 4000 rpm [29]. The Max- phase Ti_3SiC_2 material was sputtered during magnetron sputtering of Cu film. It studied by Li, and results showed that the physical and mechanical properties of Cu film were significantly improved after adding new materials [30]. Guo studied the metallographic properties and hardness of the magnetron sputtering bearing of AlSn20 material, and the experimental results reached the international advanced level like Miba bearings [31]. Song prepared AlSn20Cu thin films by magnetron sputtering. The hardness of the tested films was 120 HV, and the friction coefficient was less than 0.1 under oil lubrication [32]. Although the high hardness film improves the bearing capacity of the bearing, it also reduces the adhesion of the bearing. However, if the surface hardness of bearing is less than 50 HV will have better embedding performance [33].

The different compound films can be synthesized and synthesized because maximum as eight targets can be sputtered simultaneously from magnetron

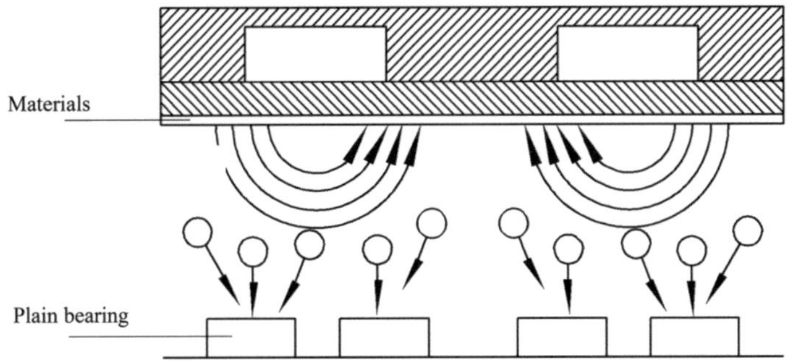

Figure 6.
Technology of magnetron sputtering.

sputtering. In the existing research, bearing alloy composite films such as Ti/Cu/N Cu_xSn_y, TiN/Cu had been prepared, and adhesin strength of these films is excellent [34, 35]. However, the tribological properties of above materials are the same as the bearing materials such as AlSn20 and AlSn20Cu. The friction coefficient is higher under dry friction conditions and mixed lubrication conditions, which cannot satisfy the self-lubricating performance requirements at wide temperature.

To improve the tribological and mechanical properties, the noble metal materials such as indium or rare earth materials are used during the electroplating plating process. However, the plating solution is a strong acid or a strong alkali substance, which is likely to cause serious environmental pollution. Magnetron sputtering equipment is expensive to manufacture. Many magnetron sputtering equipment only produce small test specimens in a laboratory environment, and cannot be mass-produced or mass-produced for large-sized sliding bearings. The utilization rate of magnetron sputtering target is generally less than 40%, and the working time is long and the production efficiency is not high when vacuuming, injection and depositing materials [36]. Therefore, it is necessary to find a new technology to prepare a self-lubricating sliding bearing.

2.2.3 Liquid dope of spraying

The most advanced surface treatment method available today is the surface spraying coating method. Compared with traditional surface treatment processes such as electroplating and magnetron sputtering, it has the advantages of environmental friendliness, high production efficiency, coating processing, and good coating lubrication performance. Spraying sliding bearings are characterized by high speed and high efficiency. It takes only several seconds to spray lubricating liquids, and large-scale and large-scale production will be realized if solidification furnaces and cooling furnaces are enough. For example, an automatic spraying production line of Shanghai Federal-Mogul company produces more than 12 million bushings with spraying coating (**Figure 7** and **Figure 9**).

The north Americans first began to study the coating technology of sliding bearing coatings and applied for related technology patents. For example, in the 1970s, Campbell used MoS_2 and Sb_2O_3 as solid fillers and epoxy resin as a binder to prepare self-lubricating coating, which was applied to sliding bearings [37]. The bearing coating prepared from materials such as MoS_2, Sb_2O_3 and epoxy resin has the advantage of low friction coefficient, but the ordinary epoxy resin working temperature generally does not exceed 140°C, which does not satisfy the long-term use requirements of temperature conditions above 200°C.

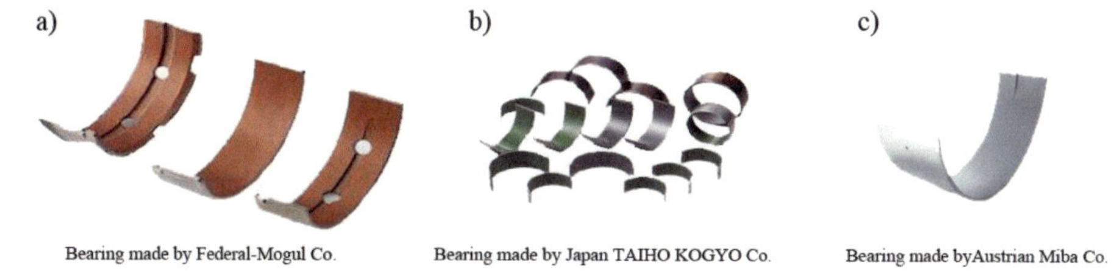

Figure 7.
Plain bearings with dope coating of overseas.

After several decades of development, the of plain bearings with self-lubricating coating made from liquid spraying have been large scale produced by developed companies such as Federal-Mogul Co., Austrian Miba Co., Japan TAIHO KOGYO Co. However, there are few research materials on self-lubricating coatings in China. In the year of 1983, the first MoS_2 self-lubricating coatings prepared by liquid dope was studied by Liu of the Institute of Coatings, however, the coatings were not applied to sliding bearings [38]. The sliding bearing surface coating technology of China started late, and it is still a new technology. In recent years, many bearing research institutes and manufacturers in China have begun to study the coating technology of sliding bearing made by liquid spraying. As shown in **Figure 8**, bearings with MoS_2 lubricating coating made by liquid spraying, and prepared with Zhejiang CSB Co. and the ZYNP Co.

The widely used coating dopes in China are MoS_2 and PTFE dopes, which wear resistance and temperature resistance are poor, and the tribological properties of composition elements such as resin and auxiliary are lower than that of Dow Corning D7409 and Kawanori of Japan. However, self-lubricating coatings for bearings under the medium (200°C) and high temperature (800°C) dopes are lacking of. The special lubricating materials of coating dopes of American D7409 and Kawanori of Japan did not particularly be selected according to the special alloys of bearing likewise the different aluminum alloys, copper alloys, and babbitt alloys, and these coatings drop easily if the temperature varies with time. Therefore, it is important to develop new coating dope for special alloys at different temperature.

The coating spraying technology includes coating dope preparation, liquid dope spraying through spraying gun and solid coating formation. The coating dope consists of self-lubricating materials, anti-wear materials, resins, auxiliaries and solvents. The different lubrications and anti-wear materials easily mixed together due to the liquid solution, and the coating made from dope will have the excellent tribological properties. The principle of paint spraying is shown in **Figure 9**. There are two ports on the spray gun connected to the spraying gas and dope respectively.

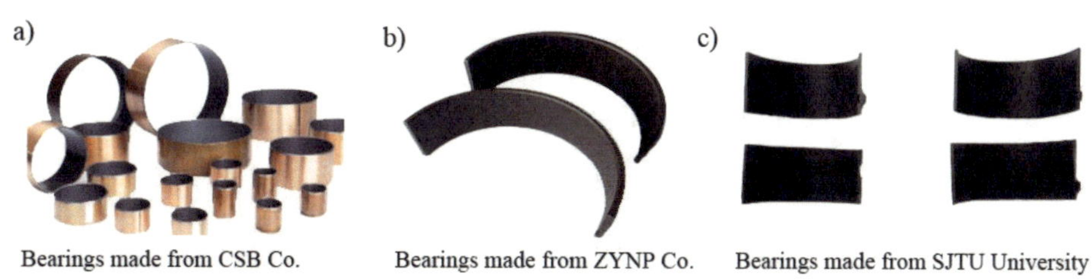

Figure 8.
Plain bearings with dope coating of China.

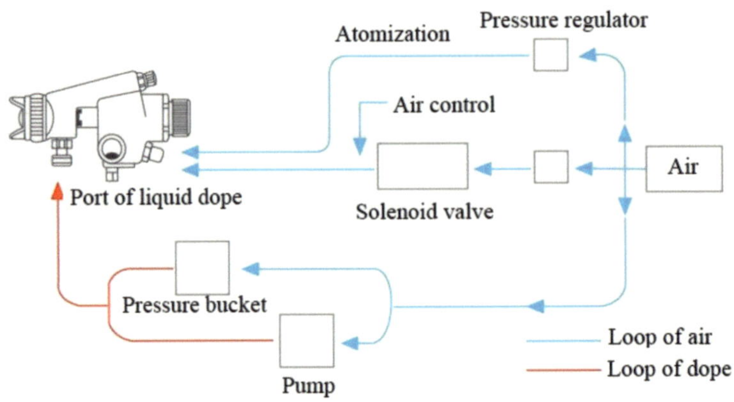

Figure 9.
Work principle of coating dope spraying

The paint dope is in a pressure tank with automatic stirring. It is driven into the spraying gun by the action of the pump to adjust the size of the gas source, liquid flow rate and spraying distance to control the amount of spraying dope.

2.2.3.1 Coating formation

Guo used epoxy resin as the binder, and MoS_2 and PTFE were used as the main lubricating materials to prepare the antifriction coating. The friction mechanical properties of the coating were studied at different curing temperatures. The results showed that the coating performance was the best when the coating formation temperature was 120°C. The coefficient of friction is 0.125, and the adhesion is 16.73 N [39]. However, ordinary epoxy resins have poor temperature resistance, and the coating thus prepared cannot be operated for a long period of time at 200° C. Cao studied the spraying distance effects on coating spraying efficiency and coating thickness uniformity. He pointed out that reducing the spraying distance improves the adhesion between the coating and the substrate. However, the shorter the distance was, the worse the coating thickness uniformity could be [40]. Yang used sagging as the object of assessment, and studied the spraying distance and spraying temperature during the PTFE coating preparation, but the principle and theory of sag phenomenon had not studied [41].

2.2.3.2 Porosity

The heating temperature of the coating to form a coating is generally from 120 to 220°C. Under the action of materials such as resin and polyimide, the liquid coating has a sealing effect on the solid coating process which leads to low coating porosity. Li uses rare earth materials and a rapid thermosetting method to prepare a coating, and the porosity of coating is just only 0.35% [42]. The porosity of sliding bearing produced by the powder metallurgy and electro-plating is high. Lins et al. studied the effect of current density on the porosity of nickel deposited with copper substrates. The porosity of nickel coatings was 6.22% according to different current magnitude tests [43]. The films prepared by magnetron sputtering are relatively dense, and the porosity is generally from 0.5 to 5% [44, 45]. If 1–2% reduction in porosity, the fatigue strength of the workpiece will increase from 10 to 30%, so low porosity is one of the necessary conditions for sliding bearings [46].

2.2.3.3 Bond strength

The initial state of the coating prepared on the surface of the bearing is liquid. In order to avoid sagging of the coating, the coating thickness should not be too thick. The optimum coating thickness of the coating method is less than 20 μm, and the infiltration method and the brushing method are prepared. The thickness of the coating should not exceed 120 μm. The bonding strength of the coating to the substrate is influenced by surface roughness of the substrate, the type of coating adhesive, the preheating and the curing temperature. Therefore, it is particularly important to study the coating forming process to improve the bonding strength of the coating. An used epoxy resin and polyvinyl butyral as binder to prepare coating with TiO_2 as filler. The bond strength of the steel matrix is from 9 to 12 MPa [47]. Mao prepared a coating of made by graphite, MoS_2, PPS (polyphenylene sulfide) and PES (polyethersulfone resin) by regression test. The optimum bonding strength of the tested coating was 42 MPa [48]. The self-lubricating coatings with polyamide-imide used as the binder, and the main lubricants were MoS_2 and PTFE. The adhesive strength studied by Song according to the GB9286-88 paint film rating test of China is poor as level 1, which does not reach the optimal bonding strength [49].

2.2.3.4 Lubrication properties

Self-lubricating coatings are mainly composed of solid lubricating fillers such as MoS_2, PTFE, WS_2, graphite, and some additives such as polyimide, epoxy resin, phenolic resin, leveling agent, dispersing agent, defoaming agent, etc. Gao used PAI (polyamide-imide polyamide-imide) as a binder, and the coating prepared by mixing 8% MoS_2, 5.4% PTFE and 1.5% graphite had the best performance. The coating had not changed which was tested at 80°C and −40°C, and it was not changed for socking in 10% HCl over 3 months [50]. Li made PTFE coating which added MoS_2 and graphite as the solid lubricants. The PTFE used as the main component, and PEEK (polyether ether ketone), PI (polyamide) and PPS as binders. The coefficient of friction of the PTFE coating was 0.12 [51]. The studies of self-lubricating sliding bearing coatings are mainly concentrated in recent years, so there are few researches of self-lubricating coatings which used at wide-temperature. Ordinary epoxy resins and polyimide materials are generally used in coatings, however, they cannot work for long under condition of 200°C. In addition, the sagging and leveling phenomenon of the sprayed sliding bearing have not been studied in detail. The theory of optimal thickness of coating, optimum roughness of the substrate, optimum curing temperature of the coating and cooling temperature of the coating are also very few.

Compared with the self-lubricating sliding bearings prepared by electroplating and magnetron sputtering, the paint-coated bearings have the advantages of environmental protection and high production efficiency. The review of coating research shows that the development of self-lubricating coatings under different temperature is one of the best technologies for preparing sliding bearings due to the excellent wear resistance and low friction performance.

2.3 Review of self-lubricating sliding bearing at high temperature

The work temperature of socket bearings in the missile launcher mechanism is higher than 800°C. Materials such as resins used as components in dope coating cannot be operated under ultra-high temperature conditions. At present, the maximum working temperature of grease-based lubricating materials generally do not exceed 200°C, and the maximum using temperature of polymer-based self-lubricating materials (single-phase) generally does not exceed 400°C [52]. The ranges of optimum lubrication for single-phase materials are shown in **Table 1**.

The frictional coefficient of MoS_2 is generally lower than 0.1 under room temperature. The oxidations of Mo element will be generated if MoS_2 working temperature exceeds 200°C, that the self-lubricity of MoS_2 is reduced. The MoS_2 lubrication performance will be further reduced and the self-lubricating effect will be lost if the temperature exceeds 350°C. In **Table 1**, B_2O_3 produced by BC_4 under high temperature has self-lubricity, and its friction coefficient is from 0.10 to 0.30, while the BC_4 friction coefficient is from 0.35 to 0.40 at room temperature [55]. Single coatings are difficult to maintain self-lubricating performance over a wide temperature range. Ouyang used $BaSO_4$, $BaCrO_4$, Ag as the main materials to prepare self-lubricating materials. The results of friction and wear tests showed that the friction coefficient was from 0.38 to 0.55 when the temperature was from room temperature to 800°C [56]. Zhen studied a self-lubricating material mainly consisted of CaF_2, BaF_2 fluoride and noble metal Ag, and the frictional coefficient of the composite material is from 0.24 to 0.3 at the temperature from room temperature to 800°C [57]. The physical and phase change will vary with temperature, for example, the structure and tribological properties of Ti_2AlC coating made by thermal spraying are different at room temperature and 800°C [58]. The coating shown in **Figure 10a** is spongy and many pores, however, the coating was oxidized and became dense at 800°C.

2.3.1 Powder metallurgy sintering

In order to make the bearings have self-lubricating performance under ultra-high temperature conditions, the common method is to prepare self-lubricating coating by powder metallurgy method. In the powder sintering method, the self-lubricating powders often mixed with some functional powders, and placed together in a high-temperature furnace to prepare a self-lubricating composite material. In the preparation of self-lubricating composites by powder metallurgy, local unevenness will be generated due to uneven powder mixing, uneven laying, uneven powder size or incomplete sintering of the powder during sintering [59].

Chen studied the uneven microstructure of AlSi alloy powder. The study showed that when the mass fraction of AlSi alloy powder is 50%, it can effectively reduce the unevenness of the product and improve the stability of the alloy [60]. Ding studied the powder metallurgy oxidation behavior at different temperatures, different processes and different atmospheres, and results pointed out that the amino atmosphere was more protective than nitrogen [61]. Cao prepared Ti_6Al_4V coating by sintering 23 μm Ti particles and 40 μm Al-V powder. The powder was cold pressed to 180 MPa before sintering, and then sintered at 1250°C in vacuum to prepare Ti_6Al_4V coating. The layer porosity was only 3.5% [62]. Studies had shown

	Graphite	MoS_2	WS_2	WSe_2
Frictional coefficient	0.05–0.15	0.05–0.20	0.08–0.20	0.09–0.20
Max using temperature	300°C	340°C	450°C	450°C
	TaS2	**PbO**	**BN**	**BC4**
Frictional coefficient	0.05–0.20	0.07–0.20	0.06–0.20	0.10–0.30
Max using temperature	550°C	700°C	900°C	1200°C

Table 1.
The best frictional coefficient of single material [53, 54].

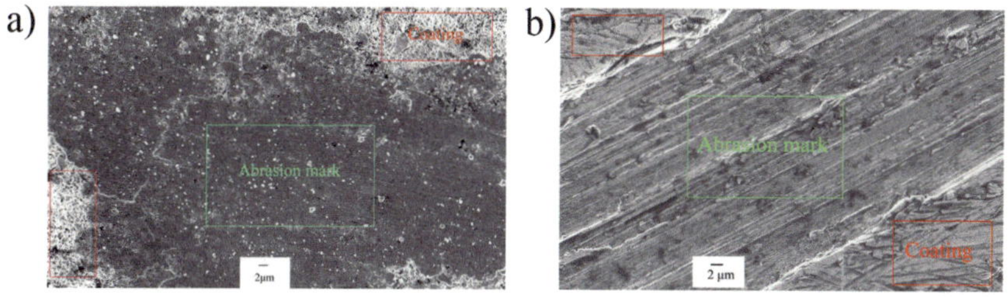

Figure 10.
Structure and abrasion marks of Ti_2AlC coating at different temperature: (a) room temperature; (b) 800°C.

that increasing pressure during powder metallurgy preparation of products, and using an atmosphere to protect the environment or vacuum environment is conducive to the reduction of porosity.

As shown in **Table 2**, the frictional coefficient of self-lubricating materials prepared by the powder sintering method is from 0.26 to 0.8 under different temperature conditions. Although self-lubricating materials at wide temperature domain had achieved some success, they had not been applied to sliding bearings. Most of the results were in the laboratory stage, and the self-lubricating composite materials prepared by powder sintering had defects such as high oxidation and high porosity. In **Table 2**, in order to reduce the high temperature frictional coefficient, a highly toxic fluoride material under high temperature conditions was used, which was not conducive to safe production.

The CoFs (coefficients of friction) of self-lubricating materials studied in **Table 2** is high and at 800°C. In order to satisfy the requirements of self-lubrication of sliding bearings used at wide temperature range, the new process and method should be used.

2.3.2 Thermal spraying

The solid self-lubricating powder particles are heated and molten by thermal spraying, and then directly sprayed onto the surface of the sliding bearing to form a coating. Since there is no resin, the prepared coating can be applied to a super-high temperature working condition. The kerosene, propane, or hydrogen used as burn gas during the supersonic flame spraying, and the molten or semi-molten alloy powders are sprayed with high speed on the surface of substrate by flame spraying. The time of alloys contact with oxygen in the air is very short due to the high spraying speed of 500 m/s. Supersonic flame spraying technology widely used as surface additive processing technology, and have high production efficiency and good coating bonding strength. A variety of coatings had been successfully prepared by thermal spraying

Materials	Lubrications	CoF at room temperature	CoF at 600°C	CoF at 800°C
Ag-Pb-Cu-Sn [63]	Ag-Pb	0.35	0.3	—
ZrO_2-MoS_2-CaF_2 [64]	MoS_2-CaF_2	0.3	0.8	0.27
Ni_3Al-Ag-Mo-BaF_2/CaF_2 [65]	Ag, BaF_2/CaF_2	0.35	0.38	0.32
Ni-Ag-BaF_2/CaF_2 [66]	Ag, BaF_2/CaF_2	0.3	0.32	0.26
NiCr/Cr_3C_2-WS_2 [67]	WS_2	0.42	0.28	—

Table 2.
CoFs of self-lubricating composite materials at high temperature.

technology, and which also can be used as sliding bearing materials. Thus, the preparation of sliding bearing coatings in combination with thermal spraying technology is reliable. Supersonic flame spraying is shown in **Figure 11**. The self-lubricating powders enter the spraying gun firstly, then, with the action of prismatic shock wave, the molten powders are sprayed on the surface of substrate to form a coating.

High bonding strength is one of the advantages of supersonic flame spraying. The bonding strength of WC-10Co-4Cr coating made by supersonic flame sprayed and prepared by Zhu was 72.63 MPa, and the porosity was 1.5% [68]. WC–Co coating studied by Tang, and prepared on the surface of the screw material substrate had a bonding strength of 65 MPa [69]. The bonding strength of $MoSi_2$ coating prepared by Wu was only 14.5 MPa, indicating that the material and spraying parameters were important factors for the bonding strength of the coating. If bonding strength is low, the bonding strength can be improved by optimizing the spraying parameters [70].

The 54% Cr_3C_2 added with 34% NiCr and 12% CaF_2/BaF_2 mixed powders, and sprayed as coating was studied by Xue. The frictional coefficient of coating was 0.75 and 0.37 at room temperature and 500°C respectively [71]. Hou used supersonic flame spraying process to prepare aluminum bronze alloy coating on steel substrate, and the frictional coefficient of coating was from 0.08 to 0.12 under different oil lubrication conditions [72]. Copper alloy and aluminum alloy materials are also suitable for supersonic spraying to prepare coatings. The studies like $AgSnO_2$/Cu coating prepared by Chong, Al-Si coating prepared by Wu, and Cu(In, Ga)(S, Se)2 coating prepared by Park [73, 74].

2.3.3 Cold spraying

Cold spray technology is a new spray technology developed from thermal spray technology in recent decades. The obvious spraying characteristics are low temperature and high speed. As shown in **Figure 12**, under the action of high pressure gas, the powder particles are supersonic flying through the Lava tube and deposited directly on the surface of the plain bearing to form a coating by the pure plastic deformation.

The sliding bearing materials are generally copper alloys, aluminum alloys and tin alloys. These metal materials have an ability of excellent plastic deformation during the formation of coating by cold spraying. Guo, Li and some scholars studied the coating properties of bearing alloy materials such as CuSn6, CuSn8, Cu5Sn95, AlSn5, AlSn10 and AlSn20 made by cold spraying, and the frictional coefficient of bearing alloy materials under dry friction conditions were generally

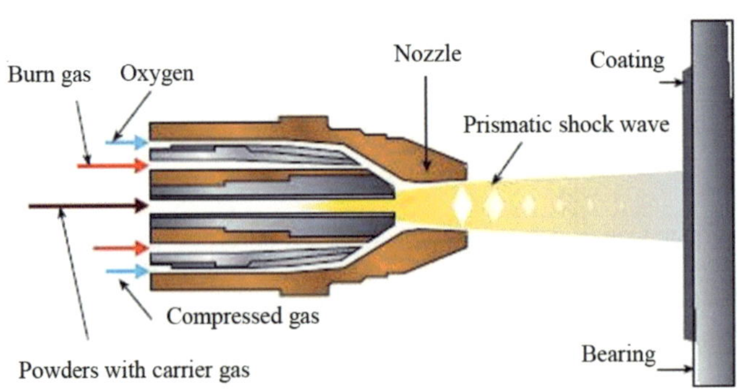

Figure 11.
Principle of supersonic flame spraying.

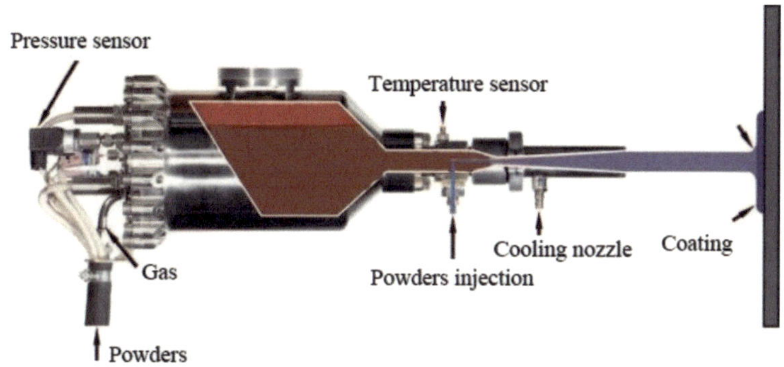

Figure 12.
Principle of supersonic cold spraying.

higher than 0.5 [75–79]. The preparation of supersonic cold spray coating is mainly based on the plastic deformation ability of powder particles, so the material should have excellent elastoplasticity [80]. The self-lubricating powder materials such as MoS_2, graphite and h-BN have poor elastoplastic deformation ability, and they are difficult to deposit on the bearing to form a coating.

The material frictional tests of coatings made by powder sintering, supersonic thermal spraying and supersonic cold spraying showed that the supersonic flame spraying and cold spraying are the desired methods to prepare coating used at high temperature. However, many materials lack of plastic deformation ability that the liquid dope spraying is the best method to deposit coating used at temperature lower than 200°C.

3. Comparisons and characteristics of different self-lubricating treatment methods

According to the studies of self-lubricating sliding bearing and self-lubricating composite materials at room temperature, medium temperature and high temperature, bearings with self-lubricating liners and bearings inlaying with solid lubricating materials are used at room temperature commonly. Bearings treated by electroplating, magnetron sputtering and liquid spraying are used at the temperature lower than 200°C. However, there are no bearings that have the continue self-lubricating performance from room temperature to 800°C, though the coating made by powder sintering, supersonic thermal spraying and cold spraying.

The production efficiency, oxidation rate, coating thickness, porosity and bonding strength compared and shown in **Table 3** according to the different surface treatments and bearing using temperature. As shown in **Table 3**, the coating thicknesses prepared by electroplating, magnetron sputtering and liquid dope spraying is thin. The electroplating is environment polluting.

The production efficiency of film produced by magnetron sputtering is low. It takes more than 20 h to prepare a film which thickness is 10 μm. The film of large sliding bearing cannot be prepared due to the structure limit. The self-lubricating liner that added on the surface of bearing has low temperature resistance due to the fail of adhesive glue and fiber braid. The structure strength is decreased as the structure of bearing changing with inlaying solid materials. The oxidation and high porosity of bearing will be generated by the method of powder sintering. The structure of materials used for thermal and cold spraying should be spherical, and the materials used for cold spraying should have plastic deformation performance.

Technologies	Production efficiency	Oxidation	Coating thickness	Porosity	Adhesive strength	Shortages
Electro-plating	Middle	Low	<15 μm	Middle	Low	Pollution
Magnetron sputtering	Low	Low	<15 μm	Low	High	Less utilization of target
Fabric liner	Low	—	—	—	High	Woven fabric invalid at high temperature
Supersonic flame spraying	High	Middle	<1 mm	Low	High	Spherical powders
Cold spraying	High	Low	>1 mm	Low	High	Powders need plastic deformation ability
Powders sintering	High	High	>1 mm	High	High	High oxidation
Inlaying lubrications	Low	Low	—	—	—	Structure strength is decreased
Liquid dope spraying	High	Low	<20 μm	Low	Middle	—

Table 3.
Comparisons of different self-lubricating methods.

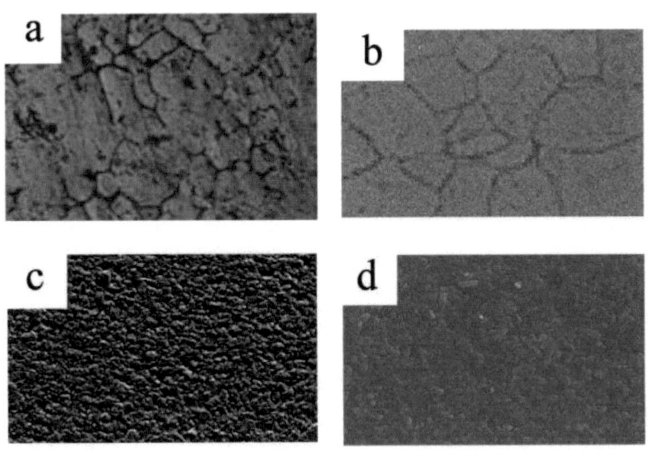

Figure 13.
Microstructure of Cu-Sn alloy of different manufacturing technologies: (a) centrifugal casting; (b) sintering; (c) electroplating; (d) magnetron sputtering.

The ceramic materials such as WC, SiC, h-BN are limit to use by the thermal spraying and cold spraying. The coating made by liquid dope spraying has the advantages of high production efficiency, green materials and good lubricity. The thickness more than 1 mm of coating on bearing prepared by supersonic flame spraying can be obtained, and it has the advantages of low porosity, good bonding strength and high production efficiency. Therefore, low- temperature and medium-temperature self-lubricating coatings should be prepared by liquid dope spray method, and ultra-high temperature wear-resistant anti-friction coatings should be prepared by thermal spraying.

The micro structure of tin and cooper coating made by different manufacturing technologies are shown in **Figure 13** [81–85]. The obvious cracks can be found in **Figure 13a, b** which manufactured by centrifugal casting and powder metallurgy sintering respectively. The coating made by magnetron sputtering is more smooth than that of electroplating, and the porosity of coating made by magnetron sputtering is little than that of electroplating. The tribological and mechanical properties of alloy coating made by centrifugal casting and powder metallurgy sintering are poor from the results shown in **Figure 13**. However, the shortages of electroplating and magnetron sputtering are the pollutions and low efficiency respectively. The new coating formation technologies such as liquid dope spraying and supersonic flame spraying should be applied for sliding bearings.

4. Current problems of self-lubricating plain bearings

The traditional self-lubricating treatments of sliding bearings are mainly electroplating, magnetron sputtering, bonding self-lubricating liners and inlaying solid self-lubricating materials. The new self-lubricating preparation technologies of sliding bearings are mainly the liquid dope spraying and supersonic flame spraying. The problems of sliding bearings are listed as follows through above investigations and discussions.

1. High friction coefficient and low fatigue life

   Most of the sliding bearings have not been surface treated, and these bearings have high dry frictional coefficients, low fatigue strength, low life and low bearing capacity. The frictional coefficient of copper alloy, aluminum alloy and tin alloy of sliding bearings is generally higher than 0.5 under dry friction condition, and it does not have self-lubricity under a wide temperature range. The traditional copper alloys of sliding bearings contain Pb metal, which is polluting and harmful to human. The self-lubricating materials at ultra-high temperature are still in the laboratory research stage, and the sliding bearing with self-lubricating cannot be produced on a large scale.

2. Environmental pollution and low production efficiency

   The usage of acid, alkali and heavy metal solutions during the production of self-lubricating sliding bearings will cause serious environmental pollution. Magnetron sputtering takes long time during vacuuming and injection processes, and the efficiency of depositing thin films is lower than other processes. The utilization efficiency of target materials of magnetron sputtering is generally less than 40%. The multiple modification treatment processes and bonding processes are required for preparing self-lubricating bearings using of bonding self-lubricating fabric liners and bearings with inlaid solid self-lubricating materials. The punching holes and inlaying solid lubricating materials are required during the production. The production efficiency of inlaying solid lubrications is low, which cannot satisfy the requirements of large-scale mass production.

3. The theory of spraying process for sliding bearing is not deep enough

The liquid dope spraying method is one of the best methods for preparing self-lubricating sliding bearings which used at room temperature and medium temperature. However, the spraying mechanisms such as leveling and sagging coating on bearing surface have not been studied. The optimum thickness of the coating, the optimum surface roughness of the substrate, the optimum curing temperature of the coating, and the optimum cooling temperature of the coating were not studied in details.

4. Poor tribological properties

The wear-resisting and self-lubricating performances of coatings are poor, and their service life under the medium temperature (200°C) is short. There are no special materials which have low difference of thermal expansion and similar physical properties with the given materials of bearings. The self-lubricating materials used at high temperature are currently in the state of laboratory research stage, and have not been prepared for bearings at large production scale.

5. Conclusion

The processing technologies and material properties of self-lubricating sliding bearings made by electroplating, magnetron sputtering, bonded self-lubricating liners, and embedded solid self-lubricating materials are studied and summarized in this paper. The advantages and disadvantages of self-lubricating bearings made by different technologies are shown. The widely surface treatment of sliding bearing is electro-plating, however, the bond strength is lower than magnetron sputtering, supersonic flame spraying, cold spraying and powder sintering technologies, and the solutions of electro-plating is pollution and harmful to human. The properties of thin film made by magnetron sputtering is excellent, however, the production efficiency is too low due to the vacuum and deposition process. The large-scale size of bearings such as bearings used in ship diesel engine cannot be deposited due to the structure limit of magnetron sputtering machine. The porosity of bearing made by powder sintering is high, and the alloys are oxidized at the high sintering temperature. The mechanical properties of cold spraying are better than thermal spraying, however, the materials of cold spraying should have excellent plastic deformation ability, that the materials such as MoS_2, C, h-BN and many ceramics materials are not able to be deposit on the surface of bearing.

Through comparative analysis, liquid dope spraying method is suggested to be adopted as the surface treatment process for bearing using at room temperature and medium temperature. The solid powder thermal spraying is suggested to be used for preparation of bearings working at high temperature. The liquid dope spraying is used in the advanced sliding bearing manufacturing companies, and the materials of liquid dope should be improved due to the wide range temperature variation at start and stop stage. The self-lubricating coating at high temperature is lacking, and the tribological properties of bearings at high temperature are poor. There were few materials that had continuous self-lubricating properties at the wide range temperature. According to the review, the lubrication materials used at high temperature mostly were the fluorides which were poisonous at high temperature. According to the summaries of self-lubricating treatments, the green materials, coating formation mechanisms, technology processes and tribological properties of liquid dope spraying and supersonic flame spraying for sliding bearings should be studied further.

Acknowledgements

This research was supported by Shanghai Xiangsheng beco engine bearing Co, Ltd. and its chairman of Shengxiang Zhu.

Author details

Jun Cao[1,2]*, Liang Qin[3], Aibing Yu[1,2], Haibo Huang[1,2], Guoping Li[1,2], Zhongwei Yin[3] and Huiyu Zhou[4]

1 School of Mechanical Engineering, Ningbo University, Ningbo, China

2 Part Rolling Key Laboratory of Zhejiang Province, Ningbo, China

3 School of Mechanical Engineering, Shanghai Jiaotong University, Shanghai, China

4 School of Natural and Applied Science, Northwestern Polytechnical University, Xi'an, China

*Address all correspondence to: caojun@nbu.edu.cn

References

[1] Zeng ZQ, Wang YK, Wang ZH. 2D simulation of the aerodynamic micro air journal bearing for micro gas turbine engine. Applied Mechanics and Materials. 2012;**121-126**:3087-3091

[2] Han QL, Li GD, Zhao WX. Bearing of high speed of rail failure analysis at rising temperature. Electric Drive for Locomotives. 2017;**1**:115-117

[3] Summer F, Grüna F, Offenbecher M, Stuart T. Challenges of friction reduction of engine plain bearings—Tackling the problem with novel bearing materials. Tribology International. 2019;**131**:238-250

[4] Sander DE, Allmaier H. Starting and stopping behavior of worn journal bearings. Tribology International. 2018;**127**:478-488

[5] Lu ZC, Zeng MQ, Gao Y, Zhu M. Significant improvement of wear properties by creating micro/nano dual-scale structure in Al–Sn alloys. Wear. 2012;**296**(1-2):469-478

[6] Liu TS, Guo YS, Song YH, Sun SW, Zhou WX. Defect analysis and elimination of $ZCuSn_{10}Zn_2$ products by centrifugal casting. Physical Testing and Chemical Analysis Part A: Physical Testing. 2013;**49**(1):52-54

[7] Novak M, Chrastny V, Sebek O, Maetinkova E, Curik J, Veselovsky F, et al. Chromium isotope fractionations resulting from electroplating, chromating and anodizing: Implications for groundwater pollution studies. Applied Geochemistry. 2017;**80**:134-142

[8] Huang SY, Qu FQ, Miao Y, Meng ZK. Development of cylindrical rotating magnetrons with high target utilization rate. Journal of Vacuum Science & Technology. 2000;**20**(2):123-125

[9] Liu HT, Li ZX, Wang JM, Sheng CX, Liu WL. Wear resistance properties reinforcement using nano-Al/Cu composite coating in sliding bearing maintenance. Journal of Nanoscience and Nanotechnology. 2018;**18**(3):2152-2157

[10] Tang ZQ, Liu XJ, Pang MH, Liu K. Surface texturing design and frictional experiment of friction pair of grease lubricated spherical plain bearings. Transactions of the Chinese Society of Agricultural Engineering. 2016;**32**(12):61-67

[11] Aderikha VN, Shapovalov VA. Mechanical and tribological behavior of PTFE–polyoxadiazole fiber composites effect of filler treatment. Wear. 2011;**271**:970-976

[12] Li HL, Zeng FP, Yin ZW, Huo YJ. A study on the tribological behavior of hybrid PTFE/Kevlar fabric composites filled with nano-SiC and/or submicron-WS_2 fillers. Polymer Composites. 2016;**37**(7):2218-2216

[13] Li YC, Qiu M, Miao YW. Bonding behavior and tribological property for self-lubricating spherical plain bearings with PTFE/Kevlar woven liners. China Mechanical Engineering. 2016;**27**(2):222-229

[14] Zhan SH, Bai YX, Qiu M. Effects of liners modification on tribological properties of self-lubricating spherical plain bearings. Bearing. 2013;(2):33-35

[15] Shen JT, Top M, Pei YT, Hosson JD. Wear and friction performance of PTFE filled epoxy composites with a high concentration of SiO_2 particles. Wear. 2015;**322-32**:171-180

[16] Zhang ZZ, Xue QJ, Liu WM, Shen WC. Effect of rare earth compounds as fillers on friction and wear behaviors of PTFE-based composites. Journal of Applied Polymer Science. 2015;**72**(3):361-369

[17] Wei Q, Zhou BY, Chen SK. Investigation on friction and wear properties of new-type inlay self-lubricating bearing materials. Material and Heat Treatment. 2009;**38**(14):81-84

[18] Zabinski JS, Donley MS, Mcdevitt NT. Mechanistic study of the synergism between Sb_2O_3 and MoS_2 lubricant systems using Raman spectroscopy. Wear. 1993;**165**(1):103-108

[19] Li JG, Yin YG, Zhang GT, Tang HY. Investigation on friction and wear properties of FeS/copper matrix composite. Bearing. 2015;**12**:34-37

[20] Wang S, Wei W. Characterization of electroplated Ni/SiC and Ni/Al_2O_3 composite coatings bearing nanoparticles. Journal of Materials Research. 2003;**18**(7):1566-1574

[21] Li WH, Zhou XY, Xu Z, Yan MJ, Wei GR. Friction behavior of brush electro-plated Ni-PTFE nano-composite coating. Tribology. 2005;**25**(6):520-524

[22] Zhang S, Li GL, Wang DH, Xu BX, Ma GZ. Tribological properties of electric bush MoS_2-C combination plating. Journal of Materials Engineering;**2013**(1):85-90

[23] Tang WH, Jiang YX, Ba JZ, Li J, Wu WH, Tang JK. Preparation of indium–tin alloy electrode by plating and its property in electroreduction of CO_2. Electroplating & Metal Finishing. 2013;**32**(6):1-4

[24] Zhou ZP, Zhao GP, Hu YH, Gan ZJ. Preliminary study of nickel-tungsten alloy plating process. Electroplating & Metal Finishing. 2010;**29**(4):5-7

[25] Klochko NP, Khrypunov GS, Volkova ND, Kopach VR, Lyubov VM, Momotenko OV, et al. Electrodeposition of Cu-ln-Se precursors for chalcopyrite solar cells. Functional Materials. 2011;**18**(3):328-334

[26] Li Y, Tao Y, Ke D, Ma YF, Han SM. Electrochemical kinetic performances of electroplating Co–Ni on La–Mg–Ni-based hydrogen storage alloys. Applied Surface Science. 2015;**357**:1714-1719

[27] Li BS, Lin A, Gan F. Improvement of stability of trivalent chromium electroplating of Ti based $IrO_2 + Ta_2O_5$ coating anodes. Rare Metals. 2006;**25**:645-649

[28] Chen DC, Zhao ZY, Zeng P, Xie GR, Zhong GM, Xu XD. Study on heat treatment and tribological properties of electro-brush plated high-tin copper–tin alloy coating. Electroplating & Metal Finishing. 2015;**34**(24):1385-1391

[29] Li CS, Wang F, Hu ZL, Sun JR, Tang H. Preparation and tribological properties research of magnetron sputtered Sn-Sb-Cu Films. Lubrication Engineering;**2013**(9):1-5

[30] Li A, Zhou Y. Joining of Ti_3SiC_2 by magnetron sputtering a layer of Cu or Zr followed by heat treating at relatively low temperatures. Journal of the American Ceramic Society. 2011;**94**(9):3072-3077

[31] Guo YJ, Li QF, Yin ST, Ren ZY, Wang J. New bearing material for magnetron sputtering coating. Journal of Harbin Engineering University. 2001;**22**(6):93-96

[32] Song H, Liu ZY, Tang WH, Yang RT, Ling GP. Effect of deposition parameter on properties of AlSn20Cu coating prepared by magnetron sputtering. Ordnance Material Science and Engineering. 2017;**40**(6):26-29

[33] Lv TS. Experiments of bearing bush. Internal Combustion Engine Parts. 1982;(3):64-69

[34] Kelly PJ, Li H, Benson PS, Whitehead KA, Verran J, Arnell RD, et al. Comparison of the tribological

and antimicrobial properties of CrN/Ag, ZrN/Ag, TiN/Ag, and TiN/Cu nanocomposite coatings. Surface & Coatings Technology. 2010;**205**(5):1606-1610

[35] Li JX, Zhang HQ, Fan AL, Tang B. Tribological properties characterization of Ti/Cu/N Thin films prepared by DC magnetron sputtering on titanium alloy. Surface & Coatings Technology. 2016;**294**:30-35

[36] Iseki T. Target utilization of planar magnetron sputtering using a rotating tilted unbalanced yoke magnet. Vacuum. 2009;**84**(2):339-347

[37] Campbell ME, Walker WD. Polyphenylene sulfide bonded film lubricant. US Patent 23337009. 1975

[38] Liu QX, Zhu EX, Ding SZ. Study on wear resistance lubricating past for the bed plate of HZt13-3 rail switches. Tribology. 1983;**3**(4):25-31

[39] Guo PR, Qiu M, Li YC, Li QL. Effect of MoS2 on tribology and adhesion properties of polytetrafluorethylene base bonded solid lubrication coating. Materials for Mechanical Engineering. 2015;**39**(7):82-85

[40] Cao XG, Su YS, Wang N. The Analysis for influence of spraying distance on utilization efficiency of coatings. Shanghai Coating. 2011;**49**(9):40-42

[41] Yang J, Zhang L, Zhang YC. Preparation and application of PTFE aqueous dispersion modified waterborne polyurethane self-lubrication coatings. China Rubber Industry. 2018;**65**(7):752-755

[42] Li YG, Rui MA. Elimination of porosity in heavily rare-earth doped sol-gel derived silicate glass films. Journal of Sol-Gel Science and Technology. 2012;**61**(2):332-339

[43] Lins VFC, Cecconello ES, Matencio T. Effect of the current density on morphology, porosity, and tribological properties of electrodeposited nickel on copper. Journal of Materials Engineering and Performance. 2008;**17**(5):741-745

[44] Yoo JH, Ahn SH, Kim JG, Lee SY. Influence of target power density and substrate bias voltage on the electrochemical properties of type 304 SS films prepared by unbalanced magnetron sputtering. Surface & Coatings Technology. 2002;**157**(1):47-54

[45] Wang CY, Ren YX, Li QX, Li Z, Wang XX, Huang XY, et al. Preparation technology of high refractive index anatase TiO_2 film by low temperature magnetron sputtering technique. Surface Technology. 2017;**46**(5):177-183

[46] WU YP, Leng YX, Gui X, Sun H, Huang N, Zhu SF, et al. Porosity ratio and anti-corrosion properties of Cr/CrN films synthesized by unbalanced magnetron Sputtering. Materials Science Forum. 2009;**610-611**:647-651

[47] An QL. Study on the formation optimization and performance of epoxy resin powder coating [master dissertation thesis]. Shandong University; 2012

[48] Mao JY, Peng RS. Study on new PPS composite coating preparation and performance. Development and Application of Materials. 2014;**29**(1):36-40

[49] Song XF, Huang ZR, Li QM. Preparation of polyamide-imide self-lubricating composite paint and evaluation of performance of composite coating made of the paint. Material Protection. 2013;**46**(3):18-21

[50] Gao JM, Li HY, Zhou L, Hou B, Hou LW. Study on preparation and anticorrosive properties of polyamide-imide special coatings. Paint and Coatings Industry. 2014;**44**(9):31-36

[51] Li K. Study on preparation and properties of PTFE lubricant coating [master dissertation thesis]. Yanshan University; 2016

[52] Chen JM, Lu XW, Li HX, Zhou HT. Progress of solid self-lubricating coating over a wide range of temperature. Tribology. 2014;**34**(5):592-600

[53] Shi MS. Solid Lubricating Materials. China: Chemical Industry Press; 2000

[54] Hu ZB, Li HJ, Fu QG, Xue H, Sun GL. Fabrication and tribological properties of B_2O_3 as friction reducing coatings for carbon-carbon composites. New Carbon Materials. 2007;**22**(2):131-134

[55] Wu F, Wang LS, Zhang JS, Fan Y, Liu BW, Gao Y. Friction and wear properties of self-mated hot-pressed boron carbide pair. Tribology. 2001;**21**(3):214-217

[56] Ouyang JH, Sasaki S, Murakami T, Umeda K. Spark-plasma-sintered ZrO(YO)-BaCrO self-lubricating composites for high temperature tribological applications. Ceramics International. 2005;**31**(4):543-553

[57] Zhen J, Li F, Zhu S. Friction and wear behavior of nickel-alloy-based high temperature self-lubricating composites against Si_3N_4 and Inconel718. Tribology International. 2014;**75**(5):1-9

[58] Cao J, Yin ZW, Li HL, Gao GY, Zhang XL. Tribological and mechanical properties of Ti_2AlC coating at room temperature and 800°C. Ceramics International. 2018;**44**:1046-1051

[59] Xu L, Wu J, Liu YY, Lei JF, Yang R. Densification and properties of Ti-5Al-2.5Sn alloy prepared by using HIP. Titanium. 2011;**28**(4):19-23

[60] Chen XB, Fan JZ, Liu YQ. Measurement of micro-nonuniformity of 50% Si/Al composites and its influence on mechanical properties. Chinese Journal of Rare Metals. 2016;**40**(10):1008-1014

[61] Ding HT. Reason and solution of oxidation and decarbonization of iron base P/M parts during sintering. The Powder Metallurgy Industry. 2008;**18**(5):36-39

[62] Cao YK, Zeng FP, Liu B, Liu Y, Lu JZ, Gan ZY, et al. Characterization of fatigue properties of powder metallurgy titanium alloy. Materials Science and Engineering A. 2016;**654**:418-425

[63] Li JL. Study on tribological properties of novel solid lubricating material and coatings at elevated temperature [doctoral dissertation thesis]. Nanjing University of Technology; 2009

[64] Kong LQ, Bi QL, Niu MY, Zhu SY, Yang J, Liu WM. High-temperature tribological behavior of ZrO_2–MoS_2–CaF_2 self-lubricating composites. Journal of the European Ceramic Society. 2013;**33**(1):51-59

[65] Niu MY, Bi QL, Yang J, Liu WM. Tribological performance of a Ni_3Al matrix self-lubricating composite coating tested from 25 to 1000°C. Surface & Coatings Technology. 2012;**206**(19-20):3938-3943

[66] Zhen JM, Li F, Zhu SY, Ma JQ, Qiao ZH, Liu WM, et al. Friction and wear behavior of nickel alloy based high temperature self-lubricating composites against Si_3N_4 and Inconel 718. Tribology International. 2014;**75**(5):1-9

[67] Yang MS, Liu XB, Fan JW, He XM, Shi SH, Fu GY, et al. Microstructure and wear behaviors of laser clad NiCr/Cr_3C_2–WS_2 high temperature self-lubricating wear-resistant composite coating. Applied Surface Science. 2012;**258**(8):3757-3762

[68] Zhu CC, Wang MY, Hou BY, Ren XJ. Application of HVOF with WC-10Co-4Cr on high sluice valve. Journal of Thermal Spray Technology. 2013;**5**(1):11-15

[69] Tang CY, Chen RF, Shen ZX, Zhu R, Yang F, Su FS. Investigation on WC-Co wearable coating on the surface of screw substrate prepared by HVOF spraying. Journal of Yangzhou University. 2008;**11**(1):34-37

[70] Wu H, Li HJ, Lei Q, et al. Effect of spraying power on microstructure and bonding strength of $MoSi_2$-based coatings prepared by supersonic plasma spraying. Applied Surface Science. 2011;**257**(13):5566-5570

[71] Xue QZ, Zhang JP, Ma L, Wang W. Study on the fluorination high velocity oxy-fuel spraying coating. New Technology & New Process. 2010;**2010**(7):80-83

[72] Hou YF, Shen CJ, Zhang H, Hu WX. The tribological performance of HVOF Tin bronze steel bimetal material. Lubrication Engineering. 2011;**36**(1):70-73

[73] Fu C, Jiang FY, Du ZM. Preparation and characterization of $AgSnO_2$/Cu contact material by supersonic plasma spraying. In: 2010 International Conference on Mechanic Automation and Control Engineering, MAC. 2010. pp. 3442-3445

[74] Wu JW, Yang JG, Fan HY, Yoon SH, Lee CH. The bond strength of Al-Si coating on mild steel by kinetic spraying deposition. Applied Surface Science. 2006;**252**(22):7809-7814

[75] Guo XP, Zhang G, Li WY, Gao Y, Liao HL, Coddet C. Investigation of the microstructure and tribological behavior of cold-sprayed tin-bronze-based composite coatings. Applied Surface Science. 2009;**255**(6):3822-3828

[76] Li JF, Agyakwa PA, Johnson CM, Zhang D, Hussain T, McCartney DG. Characterization and solderability of cold sprayed Sn–Cu coatings on Al and Cu substrates. Surface & Coatings Technology. 2010;**204**:1395-1404

[77] Guo XP, Zhang G, Li WY, Dembinski L, Gao Y, Liao HL, et al. Microstructure, microhardness and dry friction behavior of cold-sprayed tin bronze coatings. Applied Surface Science. 2007;**254**(5):1482-1488

[78] Ning XJ, Kim JH, Kim HJ, Lee CH. Characteristics and heat treatment of cold-sprayed Al–Sn binary alloy coatings. Applied Surface Science. 2009;**255**(7):3933-3939

[79] Ning XJ, Jang JH, Kim HJ, Li CJ, Lee CH. Cold spraying of Al–Sn binary alloy: Coating characteristics and particle bonding features. Surface & Coatings Technology. 2008;**202**(9):1681-1687

[80] Deforce BS, Eden TJ, Potter JK. Cold spray Al-5% Mg coatings for the corrosion protection of magnesium alloys. Journal of Thermal Spray Technology. 2011;**20**(6):1352-1358

[81] Cao J, Yin ZW, Li Hl, Gao GY. Research progresses and suggestions of manufacturing technologies of engine bearing bushes. In: IOP Conf. Series: Materials Science and Engineering. Vol. 272. 2017. p. 012005

[82] Tsega M, Kuo DH, Dejene FB. $Cu_2ZnSnSb(S, Se, Te)_4$ film formation from selenization of sputtered self-prepared single ceramic target. Thin Solid Films. 2015;**589**(8):712-717

[83] Gabbitas BL, Ariff TF, Cao P. Synthesis of pewter alloy from tin–copper–antimony powder mixtures by microwave and conventional sintering. Powder Metallurgy. 2011;**54**(4):488-496

[84] Meinshausen L, Bhassyvasantha S, Majumdar BS, et al. Influence of indium addition on whisker mitigation in electroplated tin coatings on copper substrates. Journal of Electronic Materials. 2016;**45**(1):791-801

[85] Vahda SE. Tin-copper-lead alloy produced by horizontal centrifugal casting. Archives of Foundry Engineering. 2016;**16**(1):131-137

Chapter 9

Tribology: The Tool to Design Materials for Energy-Efficient and Durable Products and Process

Amaya Igartua, Raquel Bayon, Ana Aranzabe and Javier Laucirica

Abstract

This chapter describes a summary of the main tribological achievements carried out in TEKNIKER during the last 37 years. It covers the description of commercial and newly developed tribological test benches and case studies for a wide variety of applications. The examples refer to different tribological characterization tools for material selection (e.g., composition, surface treatments, lubricants). It makes emphasis in the failure mechanisms (pitting, scuffing, abrasion, adhesion, thermal fatigue, tribocorrosion, etc.) and friction simulation of a wide range of materials (seals, textiles, steels, cast iron, light alloys, ceramic, composites), tribological systems (mechanical components, biomaterials, tribolubrication), and environments (vacuum, ultrahigh vacuum, low or high temperature, and corrosive). A huge range of new testing equipment and protocols have been developed to simulate the mentioned failure mechanisms and working environments. This knowledge will make possible, in the future, to simulate at laboratory a still wider list of tribological systems and develop new standards. Tribology will help to implement materials solutions into energy and resource efficient products and process, to reduce carbon footprint.

Keywords: tribology, tribocorrosion, lubricants, materials, pitting, scuffing, abrasion, adhesion, erosion, thermal fatigue, die soldering, vacuum, ultrahigh vacuum, tribodesorption, outgassing, high temperature, corrosive

1. Introduction

TEKNIKER is a nonprofit applied research institution located in the armoring city of Eibar specialized in manufacturing, precision engineering, and tribology, with a history of 37 years. It is a founded member of the IK4 Research Alliance, a private alliance of seven R&D centers with 1158 persons and 308 PHD, with the mission to generate, capture, and transfer scientific and technological knowledge to industries. The tribology unit from TEKNIKER is working on failure diagnosis, surface characterization, and solving tribological problems in direct collaboration with other units from TEKNIKER like surface physics and chemistry, lubricant chemistry, laser, additive and micro-nano manufacturing, maintenance, mechanical design, sensors, and robotics. TEKNIKER is a European referent in the field of tribology with participation in more than 130 projects (EU, Spanish, and regional)

IntechOpen

with more than 250 scientific contributions in congresses, journals, and books and 3 patents. It has been active in **several international working groups** and associations, in some of them, acting as Spanish representative:

- The COST 516 about tribology (1995–2000)
- The COST 532 about tribotechnology of engines and transmissions (2002–2007)
- The COST 533 on biotribology (2003–2008)
- The Virtual Tribology Institute (VTI) involving 22 centers of tribology (2005–2016)
- The steering committee member (2005–now) of the European Materials Platform (EuMaT) holding since 2017 the co-secretary
- The NANOMAT (Basque materials and nanotechnologies team (2005–2006))
- The E!-ENIWEP, the Eureka Network for wear prevention (2006–2010)
- The MATERPLAT Spanish Materials Platform (2009–now)
- The E!-SURF, the Eureka Umbrella on Surfaces (2010–2016)
- The Network 2B Funtex on Functional Textiles (2011–2014)
- The steering committee of EMIRI, the materials for Low Carbon Energy Industrial Initiative (2012–now)
- The International Tribology Council (2014–now)
- The Austrian COMET Action X-Tribology (Excellence Center for Tribology (2012–now))
- The BIC, Biobased Industries Consortium (2014–now)
- The Knowledge-based Multifunctional Materials Virtual Institute (KMM-VIN) (2016–now)
- The ECP4 Network about plastics and composites (2015–now)
- The CAPTECH Materials for Defense of the European Defence Agency (EDA) (2016–now)
- The Coordination and Support Action (CSA) MATCH, Materials Open House (2014–2017), promoted by the Alliance of Materials A4M that was created by Suschem, Nanofuture, EMRS, FEMS, EMIRI, and EUMAT
- The European Materials Modeling and European Materials Characterization Councils (EMMC-EMCC) (2017–now)
- The materials for construction team of the ECTP Construction Platform (2018–now)

- The European Pilot Projects Network (EPPN) (2017–now)
- The Basque Country representation in the materials for batteries S3P Smart Specialization Region initiative (2018–now)

TEKNIKER has been active in the **organization of Congress** (LUBMAT, IBERTRIB, COST 516, COST 532, COMADEM, EUMAT, NANOFILMS); international meetings in cooperation with MATERPLAT, EMRS, FEMS-SOCIEMAT, EMCC, and EPPN; and **courses** such Marie Curie (Oct. 2011), LUBMAT (2012, 2016, 2018), or Erasmus+ (2018).

2. Commercial equipment

The philosophy from TEKNIKER has been to buy commercial equipment when available and to develop and construct new testing benches, when they are not easily available. Thus, TEKNIKER is equipped with the **most advanced commercial equipment** covering a huge range of testing standards such as:

- FALEX **high-performance** and FALEX multispecimen test machines (ASTM G99, ASTM D3702, ASTM D5183, ASTM D2266, ASTM D4172)
- FALEX **Four-Ball** Machine to measure antiwear properties (ASTM D 2266 for greases, ASTM D 4172 for oils) and extreme pressure properties (ASTM D 2596 for greases, ASTM D 2783 and ASTM D-7421 for lubricants)
- FALEX **tapping torque test machines** (ASTM D5619)
- OPTIMOL **SRV-III machine** to measure antiwear properties (ASTM D 5707 for greases, ASTM D 6425 and DIN 51834 for lubricants) and extreme pressure properties (ASTM D 5706 for greases, ASTM D 74121 for lubricants)
- CETR UMT-3 **reciprocal tribological test**, Block on Ring Test (ASTM G77), scratch test (ASTM D7027, ASTM G171, ISO 20502), Ball on Flat (ASTM G-133), Friction dissipation (ASTM G-203), Pin Abrasion (ASTM G132), Pin on Disc (ASTM G99), Scratch hardness (ASTM G171), Thrust washer (ASTM D3702), Rolled web friction (ASTM G143), Viscoelastic properties (DIN 53513), Indentation hardness (ISO 14577), Adhesion and mechanical failures (ASTM C 1640)
- CSM **nanotribometer** (Load from 20μN-2N)
- **Twin disc** machine with Block on Ring configuration (e.g., ASTM G176, ASTM G77)
- STRAMA FZG **gear test machine**, wear (ASTM D 4998), Micropitting (FVA 54/I-IV), DGMK 575), Pitting (FVA 2/V), Scuffing (DIN 51354-1/2, ISO 14635-1, CEC-L-07-96, IP 334/93, ASTM D5182-91), grease (DIN 74), Extreme Pressure test (FVA 243)
- **Two bearing tests** by Elgeti Engineering to characterize bearings lubricated with greases and oils (DIN51819) and to develop new lubricants, coatings, rolling bearings designs, and predictive maintenance activities

- Ducom **Erosion Tester** (ASTM G-76)
- TABER **abrasion machine** (many standards, e.g., ASTM 4060 (paints), ASTM D3884 (textiles))
- **Pendulum slip** safety testing (UNE-ENV 12663, CEN/TS 16165, BS7976-2, ASTM E303, BS EN 13748-1)
- GMG **tribometer test for flooring** security (DIN 51131, CEN/TS 16165)
- Microtest **tribometer** transformed by TEKNIKER in a tribocorrosion device
- NEURTEK **Washability** Test, humid rubbing test (UNE-EN ISO 11998, DIN 53778, ASTM D 2486, ASTM D 4213)
- CORMET **stress corrosion cracking** in liquid media with electrochemical control until 300°C and autoclave to test stress corrosion at high temperature (800°C) and gaseous media (SO_2, NO_2, CO_2, O_2) (ASTM Practice G 129, ASTM Test Method E8). NACE Standards (TM0198, TM0177)
- Corrosion test salt spray (ISO 9227, ASTM B117)
- Metrohm Autolab **potentiostats,** Electrochemical corrosion (ISO 16773-1-4), liquid corrosion (ISO 2812-1, ISO 2812-2), corrosion classification (ISO 12944-6), others (ASTM G1, ASTM G3, ASTM G5, ASTM G61, ASTM G96, ASTM G102, ASTM G106)
- **Climatic chambers** (UV, temperature, humidity, condensation), (e.g., ISO 6270, ASTM D4329, ASTM D4587, ASTM D4799, ASTM D5208, ASTM G151, ASTM G154, DIN EN 12224, DIN EN 1297, DIN EN 13523-10, DIN EN ISO 4892-1, EN 927-6, ISO 11997-2, ISO 16474-3, ISO 4892-3, ISO 20340 (Off shore, Norsok), D4585)
- **INSTRON mechanical tests** at different temperatures at 1–50 kN with speed range 0.005–500 mm/min (many, e.g., tear test (ISO 13937-2:2000))
- **Compatibility of seals** with lubricants (ASTM D471)
- Complete laboratory for **paint characterization** to measure impact (e.g., ISO 6272), falling sand abrasive test (e.g., ASTM D 968, ASTM D333; ASTM D1395; ASTM D 2205), crosscut test (e.g., ASTM 3359), bend test (e.g., ISO 6860), Persoz and Kónig hardness (e.g., ISO 1522), pencil scratch test (e.g., ISO 1518-4), cupping test (e.g., EN ISO 1520), pull off test (ISO 4624; ASTM D4541), brightness loss (Crock meter)

When possible, round robin tests have been carried out (pin on disc tests, tribo-corrosion tests) with different organizations having similar commercial machines.

3. New testing benches

The differential from TEKNIKER is the **capability to develop unique tribological test machines** and new testing configurations to design a wide variety of working conditions (temperature, load, speed) in **different environments** (atmospheric, inert, vacuum, ultrahigh vacuum, corrosive, etc.) in order to cover different tribological problems to analyze the **tribomechanism taking** place on the **rubbed surfaces** (adhesion, die soldering, galling, abrasion, pitting, micropitting, fretting, tribocorrosion, tribodesorption, tribolayer formation, impact, erosion, tactile friction, corrosion, fatigue corrosion, fretting-corrosion, thermal fatigue, fading, etc.).

The objective of the measurements carried out with the tribological testing machines is to measure differences in the material behavior. For that, a reference is always used that corresponds to a material with a well-known behavior. The standard deviation of the measurement is established with the reference material. The value of the standard deviation depends on (a) the working conditions (pressure, velocity, temperature, etc.) and (b) the properties (roughness, texturing, composition, etc.). The difference between two materials is significative, if it is higher than three times the standard deviation of the measurements. In some cases, also standard protocols have been developed (e.g., tribocorrosion tests) in collaboration with ISO/CEN groups.

The **new test benches developed by TEKNIKER** have been the following:

- New atmospheric, vacuum, and ultrahigh **vacuum tribometer** with capacity to test friction and wear at different temperatures and also analyze the tribodesorption of gases, assisted by mass spectra.

- New DEMETRA machine, to measure the **outgassing** of the lubricants, polymeric materials, and coatings that need to work under vacuum conditions (ASTM E 595, ECSS-Q-ST-70-02C).

- **New drag friction** tester to measure drag friction for different applications (e.g., paints in the ships, friction losses in heat exchangers).

- **Tribocorrosimeters** under oscillatory and rotatory conditions, where it is possible to calculate mechanical wear, corrosive (chemical) wear, and the synergy between mechanical wear and chemical wear (UNE 112086, ASTM G119 and internal protocols).

- **Bearing** tests (radial and axial), to measure torque and wear (now in Mecauto-Wisco Company).

- **Real bearing friction and wear** test configuration adapted to a Twin Disc and Falex machines.

- **Ball on rod** machine to test point contact fatigue at high pressure and speeds (12000prm, reference STP 771).

- **Cylinder on cylinder** to test line contact fatigue at high pressure and speeds (reference STP 771).

- **Thermal fatigue** test bench, to simulate cycles of high temperature (up to 900°C) and low temperature (the T°C depends on cooling unit selected). One machine is available in TEKNIKER and another in AUDI.

- **Gravelometer** to test the stone-chip (abrasives, screws) resistance to impact of coatings (paints, varnish) (reference ISO 20567-1:2005). Internal protocols have been developed to test impact resistance in polymer injection molds.

- **Clutch dynamometer** test, to measure torque and wear (now in GOIZPER Company).

- **Disc brake** dynamometer test bench, to measure the friction, wear, and thermal fatigue of disc brakes (now in Edertek-Fagor Ederlan company).

- Testing configurations of **engine component** wear and friction test have been adapted to the SRV, FALEX, and CETR machines (e.g., ASTM G181-11, ASTM G206-17). Several new protocols have been developed to simulate valve stem/guide, gudgeon pin/piston, piston ring/cylinder liner, piston ring/piston, piston skirt/cylinder liner, impact of valve seat/valve head, etc.

- **New seals** test bench under reciprocating and oscillatory movement with capacity to test with different lubrication systems and different temperatures.

- **Prosthesis simulators** involving real components for evaluating tribological properties of hip prosthesis coatings and synovial fluids, adapted to FALEX and SRV machines.

- **Car wash type test,** cleaning and abrasive testing of glass (ISO 20566) and solar mirrors (internal protocols).

4. Description of the main new test benches

4.1 CATRI (CA3UHV) machine

CATRI (CA3UHV) machine has been developed by TEKNIKER to determine the friction and wear properties in atmospheric, vacuum, and ultrahigh vacuum conditions. It is also possible to measure the gas tribodesorption detected by mass spectra, during friction and wear tests. Both the force sensor and sample holder of the experimental system have been patented, PCT/ES2009/070635 and PCT/ES2010/07273, respectively [1–4]. In this machine, the patents from CSIC ES200700480 and ES2320513B1 have also been validated. This machine allows complex tribological characterization of materials and coatings in a wide range of gas pressures from atmospheric to ultrahigh vacuum and loads (0.1–20 N). This system can be used to study tribological behavior of the coatings under gradually varying gas pressure and composition, simulating the environments on the various stages of the service life of the real aerospace systems. The machine can work under ultrahigh vacuum (5×10^{-10} mbar) and inert gas in controlled atmosphere (from 10^3 to 5×10^{-9} hPa). T°C can reach up to 350°C. It can be possible to quantify the gas tribodesorption from materials and coatings (**Figure 1**).

a) 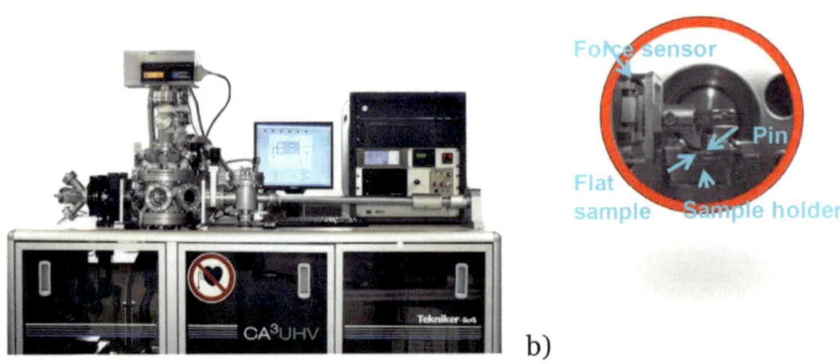 b)

Figure 1.
(a) CATRI machine, (b) detail of the sample holder, TEKNIKER©.

4.2 DEMETRA machine

DEMETRA machine is developed for the analysis of the outgassing from lubricants and materials (e.g., coatings, textiles, and plastics) in vacuum conditions. The bench test has been built using the ASTM E595-07 and ECSS-Q-ST-70-02C Reference Protocols. In the test, the volatile components of the materials and lubricants are determined at 125°C in vacuum conditions. The flexibility of the machine allows to modify working conditions. According to the standard, the Recovered mass loss (RML) should be <1% and the Collected volatile condensable material (CVCM) should be <0.1% (**Figure 2**) [3].

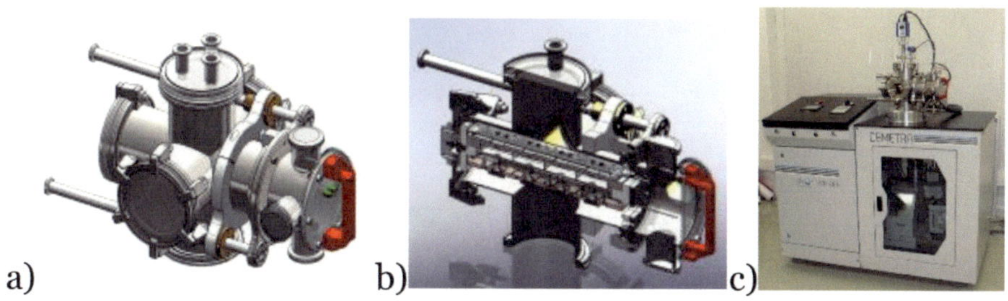

Figure 2.
Details of (a) vacuum chamber, (b) evaporation unit with four testing stations, and (c) DEMETRA machine developed by TEKNIKER.

4.3 Drag friction test

Drag friction test is for the determination of the drag friction torque of antifouling paints or surface treatments (e.g., texturing). The test samples are hollowed cylinders of 200 × 200 m that can be either located in ships or in the port to be covered with fouling after a specific time period. The torque before and after fouling and with different surface treatments can be compared at different rotating speeds, simulating either laminar or turbulent flow. Two test configurations are available with gaps of 5 and 10 mm. The machine has been developed by TEKNIKER in the frame of EU Project Foul X PEL (https://cordis.europa.eu/project/rcn/101484/reporting/en) (**Figure 3**) [5, 6].

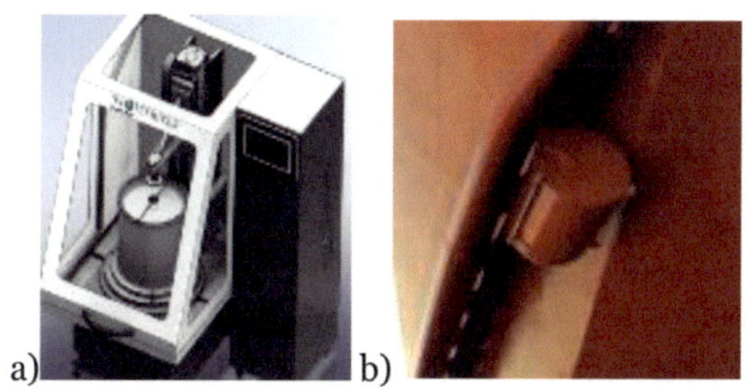

Figure 3.
(a) Drag friction tester, TEKNIKER©. (b) The image shows the testing cylinder location in a ship.

4.4 Tribocorrosion testers

Tribocorrosion testers have been developed combining a tribometer and a potentiostat. The test equipment allows monitoring and control both mechanical and electrochemical parameters. A tribometer creates relative motion, either unidirectional (rotatory) or bidirectional (reciprocating), rubbing two surfaces against each other. The electrochemical cells are used to record and control the electrochemical parameters. These cells are usually composed of three electrodes: a reference electrode, a counter-electrode, and the working electrode. The reference electrode has a stable, well-defined potential, and it is used to register the potential of the working electrode, i.e., the test sample. Typical reference electrodes are saturated calomel electrodes (SCE) and silver/silver chloride electrodes (Ag/AgCl). The counter-electrode is used to measure or control the current and is usually made of inert materials such as platinum, gold, or graphite. The electrodes are connected to a potentiostat to register the potential between the reference electrode and the working electrode or the current between the counter-electrode and the working electrode. A typical tribocorrosion test setup is schematically shown in **Figure 4** [7–16].

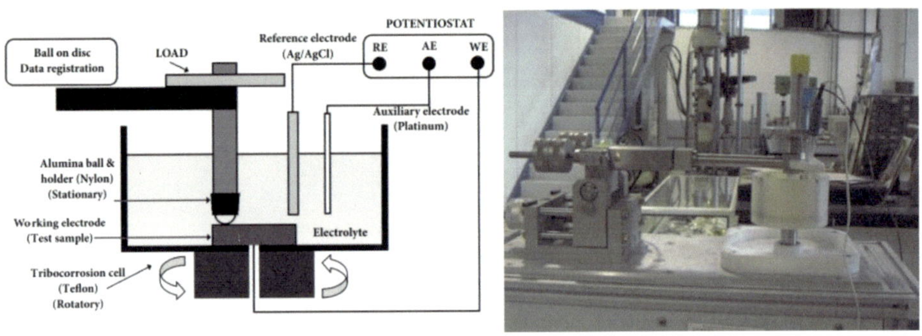

Figure 4.
Experimental setup of a tribocorrosion device for a unidirectional tribometer under ball on disc configuration, TEKNIKER©.

4.5 Test bench for plain bearings

A test bench is built with the objective to carry out friction test and wear test in plain bearings. Alternative movement has a stroke of 120–125 mm, a speed of 1–40 m/min, and a maximum load of 3000 kgf (**Figure 5**) [17, 18].

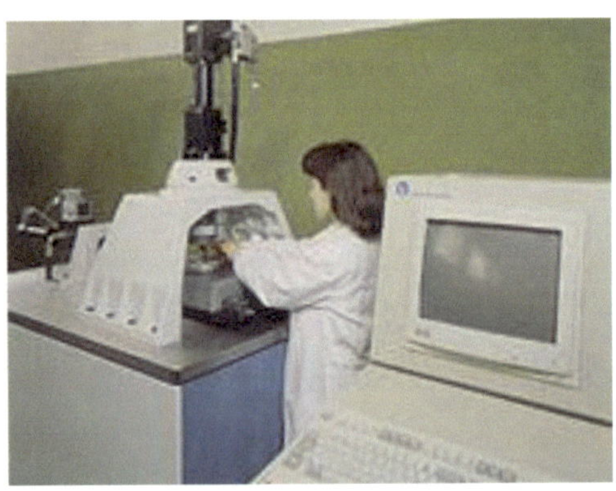

Figure 5.
Plain bearing testing machine, designed and constructed by TEKNIKER for Mecauto-Wisco company.

4.6 Ball on rod machine

Ball on rod machine is built with the objective to study the point rolling contact fatigue of the bearing materials and coatings, to apply to ball rolling bearings. It follows the protocol from Douglas Glover, "A Ball-Rod Rolling Contact Fatigue Tester" (ASTM STP 771, J. J. C. Hoo, D., American Society for Testing and Materials, 1982, pp. 107–124). It has a variable speed, reaching up to 12,000 rpm instead of 3000 rpm of the documented protocol. The typical pressure applied is 5.5 MPa. The lubrication is an air/oil system with controlled flow, typical lubrication regime for high-speed rolling bearings. Alternatively, other lubrication systems can be selected. The balls can be made from steel and ceramic and with or without coatings with a diameter of 12.7 mm. The roller has a diameter of 10 mm and length of 125 mm. The ball on rod machine has a double testing device for parallel testing. The time to reach the micropitting or pitting is recorded and monitored by an increase of the vibrations. The machine was developed by TEKNIKER in collaboration with SNR, in the frame of EU Project Eurobearing (https://cordis.europa.eu/project/rcn/6657/factsheet/es) (**Figure 6**) [19].

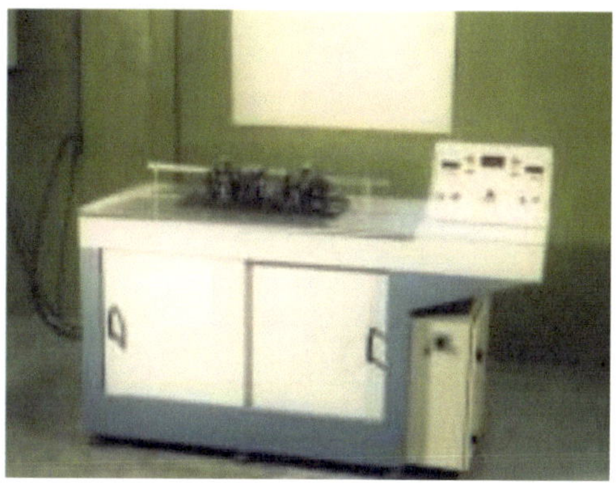

Figure 6.
Ball on rod machine, TEKNIKER©.

4.7 Cylinder on cylinder machine

Cylinder on cylinder machine, built to study the pitting phenomena in roller bearings, simulates the line contact rolling fatigue behavior of the materials. The machine is based on ASTM STP 771 protocol, with the advantage of reaching a maximum speed of 12,000 rpm with air/oil lubrication. The roller's size is 12x12mm. The duration of the test until reaching the micropitting or pitting phenomena is recorded for different types of materials and coatings of the rollers. The test stops, when the level of vibration increases, due to generation of pitting in the contact. The machine was developed by TEKNIKER in collaboration with SKF, in the frame of EU Project REFINE (**Figure 7**) (https://cordis.europa.eu/project/rcn/36428/factsheet/en).

a)  b) 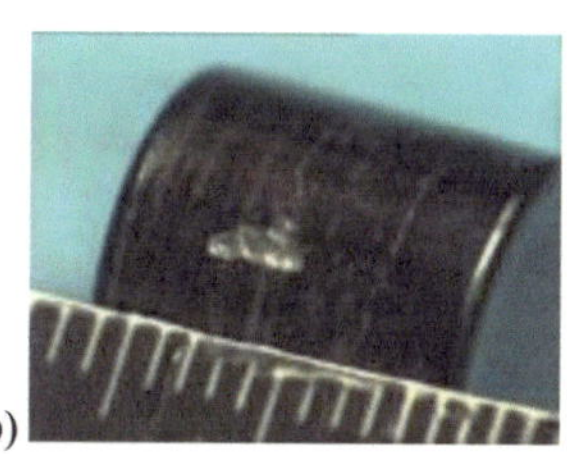

Figure 7.
(a) Cylinder on cylinder machine, TEKNIKER©, (b) pitting in a roller.

4.8 The thermal fatigue

The thermal fatigue test bench has been built to simulate thermal cycles of high temperature generated by induction heating and low temperature cooled with spray. There is one machine in TEKNIKER and one in AUDI. The machine has been developed to simulate the heating and cooling cycles of the high pressure die casting dies. The time to thermal fatigue crack failure mainly depends on (a) the maximum temperature, (b) the temperature gradient, (c) the geometry, and (d) the designed cooling channels inside the mold (testing block). The machine can simulate a time to thermal fatigue failure similar to the real applications (**Figure 8**) [20].

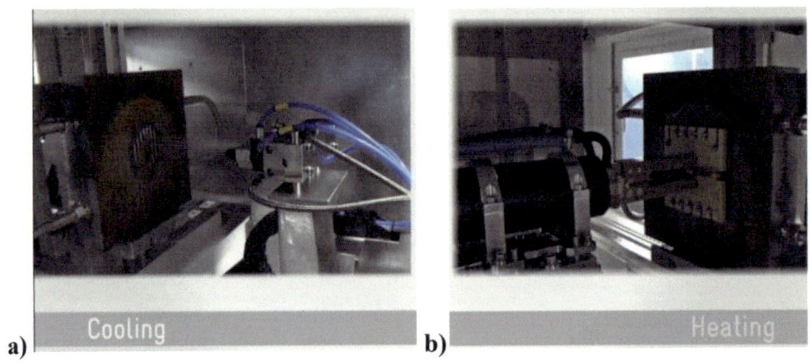

Figure 8.
(a) Cooling process, (b) heating process of the thermal fatigue machine TEKNIKER©.

4.9 Brake disc testing machine

Brake disc testing machine is built with the objective of carrying out friction, wear, and thermal fatigue tests of disc brakes and brake drums. The maximum velocity is 2500 rpm and the maximum load is 105 kg/m^2. During the test, the torque and the temperature and wear are recorded. Different protocols from automotive suppliers can be applied (e.g., AUDI, Volkswagen) (**Figure 9**) [21].

Figure 9.
Disc brake testing machine, designed and constructed by TEKNIKER for Fagor Ederlan company.

4.10 Clutch dynamometer test

Clutch dynamometer test has been built with the objective to measure the friction and wear of the clutches at fixed or variable conditions of load and speed.

Measurement of force, the friction loss between the cylinder and braking films, and temperatures at different points, forces, torque, and speeds are recorded. The specific power, specific energy, the dynamic and static torques, the sliding time, and initial speed can be monitored during breaking or evaluated during the test (**Figure 10**) [22].

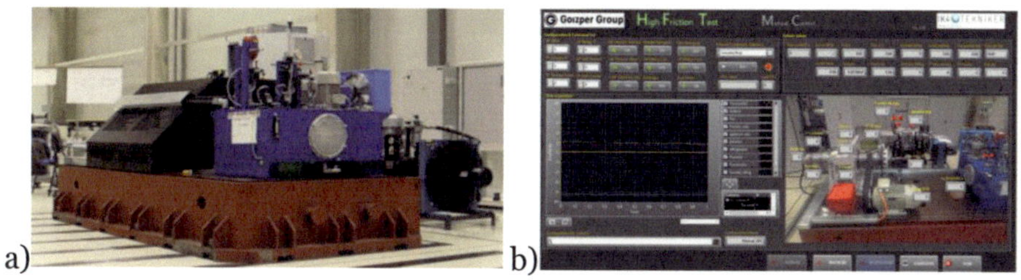

Figure 10.
(a) Clutch dynamometer test bench. (b) Control unit of the dynamometer, constructed by TEKNIKER for GOIZPER company.

4.11 The TESSA machine

The TESSA machine has been developed by TEKNIKER to measure the friction, wear, and leakage of the dynamic seals [23–28]. It can reach a maximum speed of 6 m/s. The machine has been adapted to work under variable temperature integrating a climatic chamber. The test chamber is modular, and it has the capability to measure the differential pressure between the two sides of the seals for pistons and rollers and evaluate the leakage. Oils sensors, developed by TEKNIKER (www.tekniker.es/en/ and ATTEN2 (https://atten2.com/en/), can be used to monitor the generation of wear particles in lubricated seals (**Figure 11**).

a)

b)

TEST CONDITIONS	Friction test	Accelerated life test
Lubricant pressure - p	0.3 bar	0.3 bar
Velocity - n	60-120-240-480 rpm	1000 rpm
Time	1 minute	8h/day until leakage
Temperature - T	Ambient temperature	Ambient temperature
Lubrication	Lubricated contact	Lubricated contact
Stress parameter	-	Higher pressure/rod misalignement

Figure 11.
(a) New seal test bench constructed by TEKNIKER with two units with oscillatory and reciprocal movements and climatic temperature control (low and high temperatures). (b) Example of test conditions.

4.12 Glass and mirror wash test

Glass and mirror wash test has been built to evaluate the scratch abrasion resistance of coatings used in glass of the cars and solar mirrors. In automotive sector, the glass car wash resistance is assessed. A machine-based washing is simulated in the laboratory environment using a rotating brush and synthetic dirt. The test conditions have been designed to be as close as possible to the real conditions in a car wash and solar mirrors, in order to evaluate the cleanability, the abrasion resistance, and their durability. Protocols for solar mirror evaluation are under development by TEKNIKER in the frame of EU Project In Power (http://in-power-project.eu/) (**Figure 12**).

Figure 12.
Glass and solar mirror testing device TEKNIKER©.

5. Case studies

Different case studies have been selected to cover examples of different tribological characterization tools and equipment used for material selection (composition, topography, surface treatments, lubricants), making emphasis in the failure mechanism and friction simulation of a wide variety of materials, tribological systems, and environments.

5.1 Tribology for aggressive environments

5.1.1 Vacuum and ultrahigh vacuum tribology

The **CATRI tribometer** (see **Figure 1**) has been used to screen the tribological properties under vacuum and ultrahigh vacuum of the steel/coating and titanium materials lubricated with different fluids (oils, greases, or ionic liquids). For example, it has been observed that it is possible to find non-halogen ionic liquids to lubricate steel under vacuum and ultrahigh vacuum conditions (e.g., phosphonium phosphate) and to lubricate titanium; up to now, better results have been obtained with halogen-containing ionic liquids or lubricants (e.g., $[Bu_3MeP][Tf_2N]$). The desorption of CF_3 detected in the mass spectra occurred just in the moment when the friction was reduced (see **Figure 13a**). During the test, a TiFx tribolayer was generated. Titanium and inorganic F and S(II) were detected by XPS. The reaction mechanism that explained the TiFx layer generation was elucidated (see **Figure 13c**) thanks to the fragments of the mass spectra (see **Figure 13b**) detected during the tribodesorption process. The machine and protocols have been used in the National Project CATRI [1], the EU Project MINILUBES [2], the Austrian X-Tribology COMET Initiative [3], the Eurostars Project VACUUM DLC [4], and industrial projects.

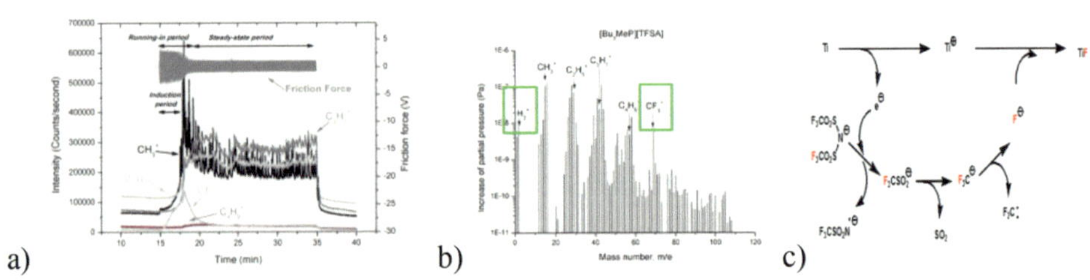

Figure 13.
(a) Friction coefficient during ultrahigh vacuum tribotest (b) mass spectra, highlighting the hydrogen and CF_3 fragments from the F-containing lubricant. (c) Proposed tribomechanism for the TiFx generation.

5.1.2 The DEMETRA machine

The DEMETRA machine (see **Figure 2**) was built to measure the **outgassing under vacuum** to screen and select the materials and lubricants that can work under these conditions. The volatility data of the outgassing measurements of the ionic liquid [Bu_3MeP][Tf_2N] has been compared to the reference lubricant, a perfluoropolyether (PFPE). It can be stated that the ionic liquid exhibits acceptable outgassing data according to the requirements of the standard ASTM E595 (RML <1.0% and CVCM <0.1%) [3]. The machine has been used in EU Projects VACUUM DLC and Austrian X-Tribology COMET Initiative. It has also been used to study dependence of outgassing of thermal fluids in function of the viscosity for heat exchangers (see Section 11.2) in EU Project SUSPIRE (http://suspire-h2020.eu/) (**Figure 14**).

Lubricants	Total mass loss (TML) %	Recovered mass loss (RML) %	Collected volatile condensable material (CVCM) %
PFPE	0.15	0.01	<0,1%
Bu_3MeP Tf_2N	<0.1	<0.1	<0,1%

Figure 14.
Outgassing data of the PFPE oil and the ionic liquid (ASTM E595).

5.1.3 Drag friction tribology in marine environments

The development of the **drag friction test** (see **Figure 3**) has been carried out in the frame of EU Project FOUL X PEL [5, 6]. In this study, the torque of different antifouling paints before and after immersion in a ship or in the Motrico Port has been compared. The tests have been carried out with artificial sea water fluid (ASTM D1141) in a container under rotation at different speed, generating laminar or turbulent regime, using different gaps of 5 and 10 mm. The effect of biofouling, roughness, and fluid properties can be measured. In the example, two fluids FX1 and FX2 were tested and being selected FX2 due to their lower drag friction in all the range of speeds tested. The machine has also been used in the EU Project SUSPIRE to test thermal fluids with different viscosities to select those with lower drag friction loss and higher energy efficiency (see Section 11.2) (**Figure 15**).

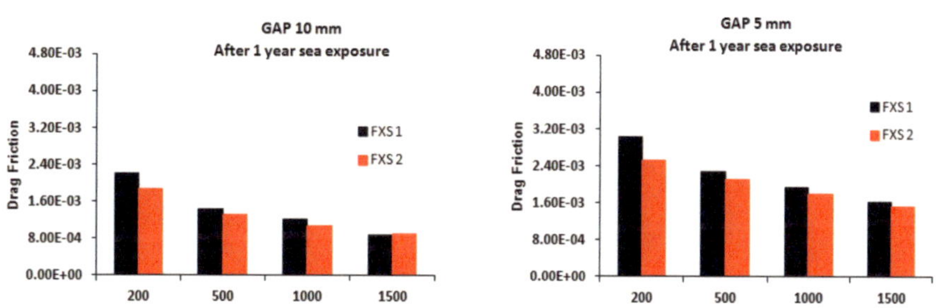

Figure 15.
Comparative drag friction measurement of two antifouling coatings, before and after 1-year sea exposure in a ship.

5.1.4 Tribocorrosion

The coupling of standard tribometers with a potentiostat has been done to measure **wear-corrosion** behaviour, that means to study the mechanical, the chemical wear and the synergy between both phenomena. In the frame of RAMPE Project [11], the tribocorrosion properties of Inconel + Cr electrodeposited coating is compared with CrN+TiN coating deposited by Physical Vapour Deposition (PVD) developed by TEKNIKER. It is observed that the impedance (corrosion resistance) after the wear test, is considerably reduced for Inconel + Cr Coating. In case of CrN+TiN PVD coating, the impedance curve resulted unaltered after wear test. These results correlate well with the wear scars measured after the test, for both coatings (see **Figure 16**). The technique has been used also in EU Projects FUNCOAT [7, 29–31], FRONTIERS [10], NADIA [32], in Regional project TRICONDEX [8], FRONTIERS [12], and National project NANOTRIBOCOR [9].

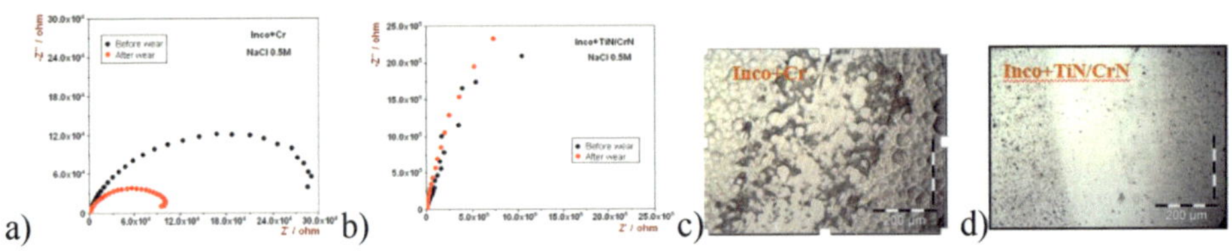

Figure 16.
Impedance measurements before and after wear test for (a) Inconel + Cr coating and (b) Inconel + PVD TiN + CrN coating. Wear scars: (c) Inconel + Cr and (d) Inconel + TiN/CrN coating.

5.1.5 Tribocorrosion in marine and offshore environments

High-strength low-alloy (HSLA) steels employed in offshore applications suffer high degradation due to their high corrosivity in seawater and the wear generated during mooring chain movement. The results, obtained in the frame of the FONDEO Adhoc project financed by the Spanish Minister of Science, showed an acceleration of corrosion during wear experiments, as consequence of the synergism generated between wear and corrosion. In **Figure 17(*i*)**, it can be seen, how the sliding has modified the electrochemical behavior of both steels. The effect of several parameters such as seawater temperature or salinity have also been studied to understand their

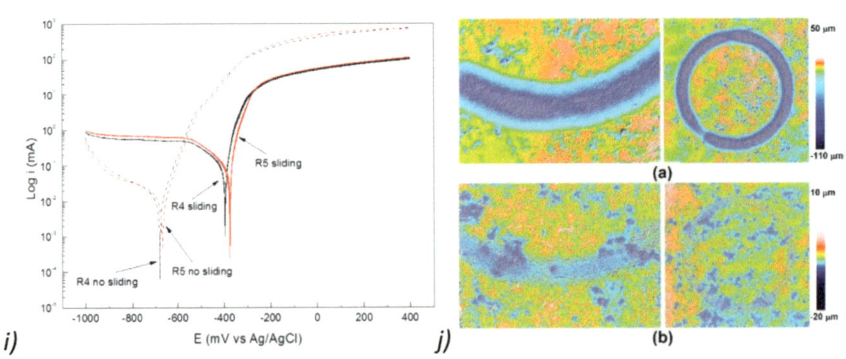

Figure 17.
(i) Potentiodynamic polarization curves obtained for two HSLA steel grades (R4 and R5) under corrosion (without sliding) and under tribocorrosion (with sliding) in synthetic seawater. (j) Topography of the wear tracks obtained after tribocorrosion tests on: (a) the TSA and (b) PEO-treated TSA.

influence in the steel degradation [13, 14]. In order to protect steels in mooring lines of floating structures, it has been developed some surface treatments based on thermally sprayed aluminum (TSA) coating with and without post-treatment by plasma electrolytic oxidation (PEO). In **Figure 17(*j*)**, it can be highlighted the reduction of wear scar after tribocorrosion tests, can be highlighted and produced when the TSA coating on the steel is additionally treated by plasma electrooxidation (**Figure 17(*jb*)**) [15, 16].

5.2. Tribology of mechanical components

5.2.1 Tribology of journal bearings

Two testing benches have been constructed (see **Figure 4**) for Mecauto company to study the torque and wear in axial and radial bearings. Also, a standard tribotest has been adapted to study the limits of pressure and velocity of the tribological pair. In the frame of BELEADFREE Project, a test has been designed to evaluate the tribological properties of real bearings. In the frame of EREBIO EU Project, the reduction of friction and contact temperature has been proven when using a biodegradable low-viscosity and environmentally friendly engine oil for the lubrication of bearings using a basic thrust washer configuration [17]. In the project ECOBEARINGS [18] and BELEADFREE, the lead content of the bearings have been reduced or eliminated (**Figure 18**).

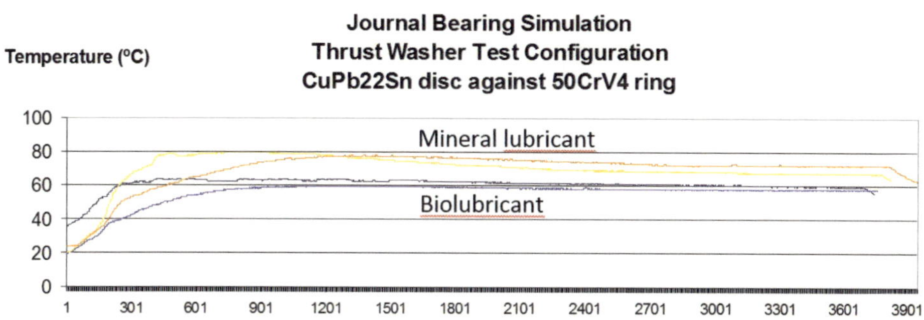

Figure 18.
Evolution of temperature when using biodegradable and mineral lubricants to lubricate journal bearings.

5.2.2 Tribology of rolling bearings

As described before, the ball on rod (**Figure 6**) and cylinder on cylinder machines (**Figure 7**) have been constructed by TEKNIKER to study point and line contact rolling fatigue to understand the pitting failure behavior of rolling bearing materials, under high contact loads and high speeds. In the frame of EUROBEARING project, different PVD coatings were studied, measuring the load-carrying capacity, the adherence, the abrasion resistance, and the hardness, as shown in the next table. Their fatigue resistance was also tested using the ball on rod machine. The long fatigue lifetime (1000 million cycles duration) of the TiN + C coating and their good behavior against abrasion and adhesion make this coating a promising alternative to be applied in high-speed rolling bearings [19] (**Table 1**). In this subject, (a) the relationship between vibration generation and oil particle generation (EU Projects TESS) [33], (b) the fatigue behavior of roller bearings (REFINE), and (c) the behavior of different PVD coatings and lubricants for rolling bearings (National Project HIEFBE) have been studied, and (d) a test configuration to study the tribology of real bearings has been developed, adapted to FALEX tribological test (Project BIOMON).

Coating type	Load-carrying capacity	Adherence	Wear protection	Hardness	Fatigue resistance
CrN	Good	Good	Good	Medium	Bad
Mo	Good	Good	Bad	Low	Bad
TiAlN	Good	Medium–good	Good	High	Good
TiCN	Bad-medium	Bad	Good	High	Good
TiAlCN	Medium	Medium	Good	High	Good
TiN + C (C sputtered)	**Medium–good**	**Good**	**Medium–good**	**Medium–high**	**Very good**

Table 1.
Qualitative behavior of coatings to increase lifetime of rolling bearings [19].

5.2.3 Tribology of clutches and brakes

Brakes and clutches require high stable friction coefficient. To understand the tribology phenomena of these systems, testing protocols have been developed for the standard FALEX (high-performance and multispecimen) tribometers to study the friction wear mechanism of brakes and clutches (high temperature abrasion, adhesion, and fading) and to study the relationship of the friction coefficient with the squeal generation. Additionally, a dynamometer test bench has been constructed to test disc brakes (**Figure 9**) and clutches (**Figure 10**). In the frame of the i-SINTER Project, the correlation between the friction measured in pin on disc tests and the torque measured in clutch tests has been studied, observing very similar values, especially at low loads (**Figure 19**). These conditions were selected for screening of the materials with pin on disc tests, before final validation in the clutch test. The tribological knowledge on dry friction has been used in the projects M-eranet JOLIE (brakes) [21], Manunet BRAKE SQUEAL (clutches) [22], and Industrial Project I-SINTER (clutches).

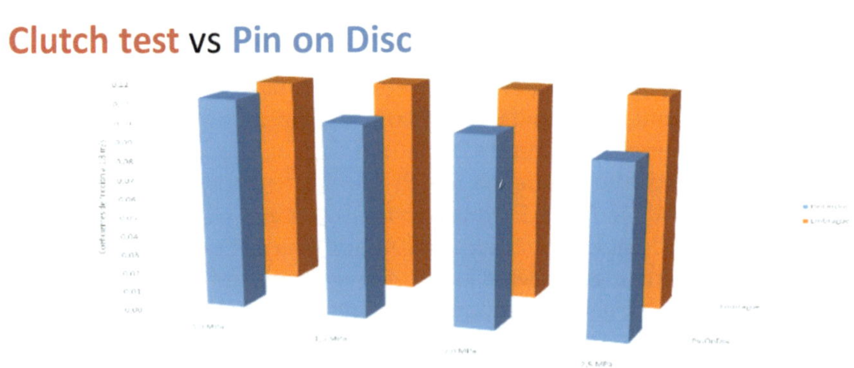

Figure 19.
Coefficient of friction during clutch test and pin on disc test carried out with the same materials.

5.2.4 Tribology of gears and transmissions

In the frame of the EU Project OPTIMIZE, TEKNIKER has upgraded the FZG machine, to measure the vibrations, noise, and transmission losses and to complement the study of micropitting, pitting, or scuffing phenomena (**Figure 20**). In this topic, the team has been working in the EU Projects TESS [33], OPTIMIZE [34], EREBIO [35–37], VOSOLUB [38], SUNOIL [39], LUBRICOAT [40], BIOGREASE [39], and BIOMON [41] and the industrial projects AEROHUMs and SELENA.

a)

b)

Sensors	Signals collected
Acelerómeters	Vibration (X, Y, Z)
Transducer in the axis of the Wheel and in the axis of the pinion	Transmission error
Torque sensor at the axis entry and in the Wheel axis.	Power loss
Microphone	Noise
Termocouple PT-100	Temperature
Drive motor	Intensity

Figure 20.
a) Upgrading of FZG machine to monitor vibration, transmission error, power losses, noise, temperature, and drive motor intensity and (b) sensors and signals collected [34].

5.3 High-temperature tribology

5.3.1 Tribology of high-pressure die casting (HPDC) and polymer injection molds (PIM)

In the frame of MUSIC Project, a thermal fatigue machine has been constructed to reproduce the main failure of the high-pressure die casting molds. In **Figure 21a**, the thermal cycles can be seen in high-pressure die casting process reproduced in the thermal fatigue machine, and in **Figure 21b**, the thermal fatigue mechanism can be observed, reproduced in a flat specimen with internal cooling circuits, representing the die. The thermal fatigue mechanism can also be reproduced in 3D complex geometries. Setup of testing protocols has been carried out to simulate also other failure mechanisms of HPDC and PIM molds such as mechanical and chemical die soldering, abrasion and erosion resistance, and the corrosion behavior. The knowledge generated in MUSIC project has been applied to the EUROSTAR Project

a) *b)*
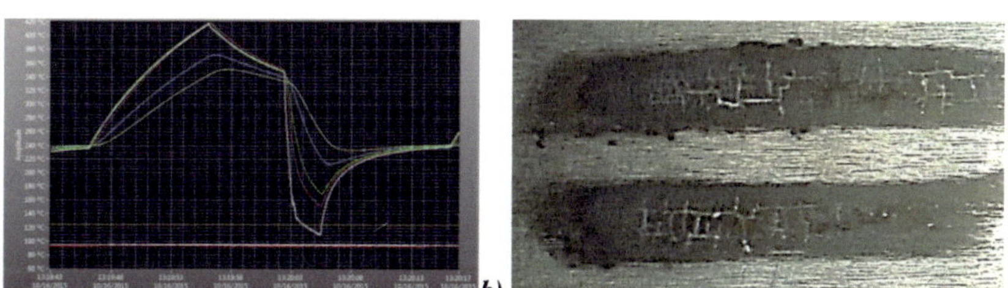

Figure 21.
(a) Thermal cycles in the thermal fatigue machine TEKNIKER©, (b) thermal cracks reproduced in the tested block of the mold material after 10,000 thermal cycles.

SUPERSLIP (to select surface treatments for polymer injection molds) and to the industrial projects HARCO, ALEPRE, MEFOLUB and MMPUL [20, 42–45].

5.3.2 Tribology of engine and turbine components

The development of testing configurations adapted to standard tribological testing machines can be highlighted in order to test real parts of engine and turbine components (e.g., valve/guide contact, piston ring/cylinder liner contact, the piston skirt/cylinder liner contact, valve head/valve inserts, gudgeon pin/piston, gland/piston rod, piston rod/seal, turbine support, etc.). In **Figure 22**, the results of scuffing tests of the piston ring/cylinder liner used to screen suitable biodegradable oils compatible with bioethanol for two stroke engine oils can be seen. Most of the lubricants suffered scuffing, except the one called SEMO36 that was suitable for the application. The testing protocols developed has been used in the EU Projects POWERFUL [46], EREBIO [47, 48], Nano-HVOF [49], EFCAP [50], COST [51], CLEANENGINE [52, 53], NANOMAG, and NADIA [32]; in the regional project TRIBORE and MOTOLURE [47]; in the industrial projects REMTRAL [51], AUMORE, RECOLURE, EQUIMOTOR and EQUIMOTOR PLUS [54], ABADIE, and 2 MW engine; and in the national projects RAMPE and ALELLA.

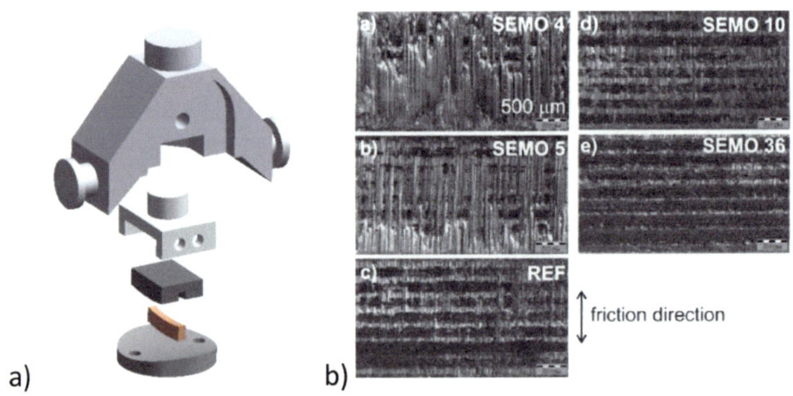

Figure 22.
(a) Piston ring/cylinder liner configuration developed by TEKNIKER. (b) Selection of SEMO 36 oil to avoid scuffing of bioethanol/biolubricant mixture, in piston ring/cylinder liner tests.

5.4 Tribology of special materials

5.4.1 Tribology of seals and elastomers

As already described (see **Figure 11**), a test bench has been constructed to test real seals to study their friction, wear, and leakage under oscillatory and reciprocal motions. A simplified tribological test has also been developed to reproduce the failure mechanism and the friction and wear of the seals. In the frame of EU Project KRISTAL, the laser texturing was applied by TEKNIKER directly on seals materials or transferred through the molds. In **Figure 23a**, it can be observed the effect of surface laser texturing on seals materials reducing significantly the friction coefficient. In the EU project STOKES, different surface texturing geometries were tested in the seals test bench constructed by TEKNIKER, selecting also the most appropriate texturing geometry. This knowledge has been applied to EU Projects FUNDTRIBO [26], ISSELUB, the regional projects MODELOST [23] and EMAITEK [24, 25], and the industrial projects 3D FLEX and ELASWEAR.

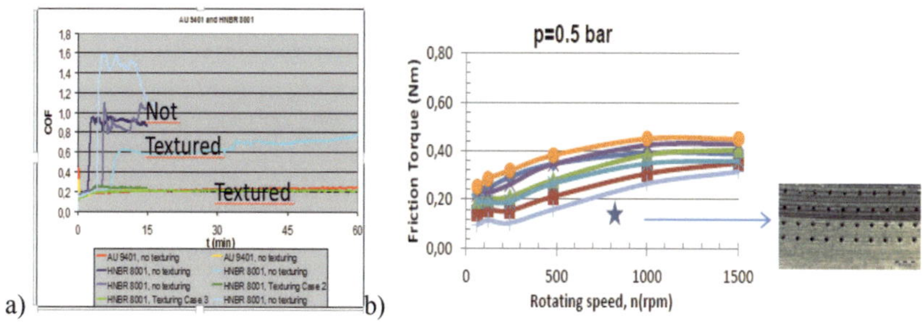

Figure 23.
(a) Effect of texturing to improve the tribological properties of different types of seals. (b) Results of the TESSA seal test bench TEKNIKER© with different texturing geometries and selection of the one reaching a minimum torque.

5.4.2 Tribology of anti-slippery floorings

In this context a test protocol has been developed to evaluate the friction (static and dynamic) and wear of the floorings determining the anti-slippery properties simulating the lifetime of the floor (**Figure 24**). This knowledge has been used in the national project TRIBOSTAND, in the EU Project SLIP SAFE, and in a regional project SEGURPAV [55].

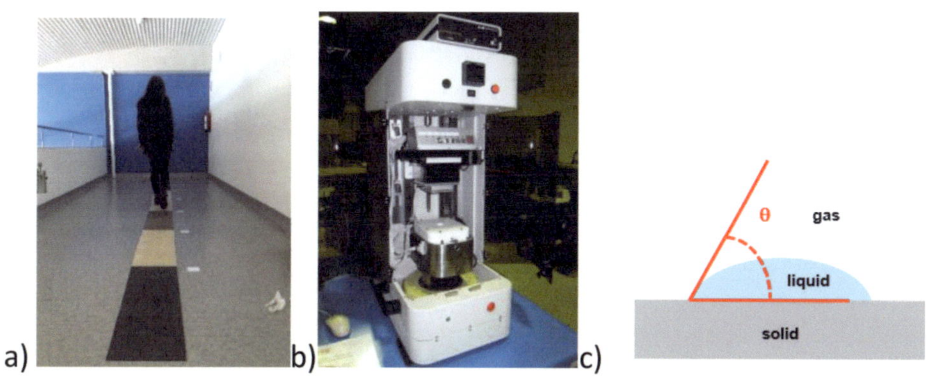

Figure 24.
Test protocols have been developed to evaluate the durability, the cleanliness, and the comfort of the floorings (a) involving persons, (b) tribometers, and (c) contact angle measurements.

5.4.3 Tribology of textiles

Testing protocols have been set up to evaluate mechanical properties, abrasion, and tactile properties of the textiles (**Figure 25**). This knowledge has been used in the EU Project 2B FUNTEX [56].

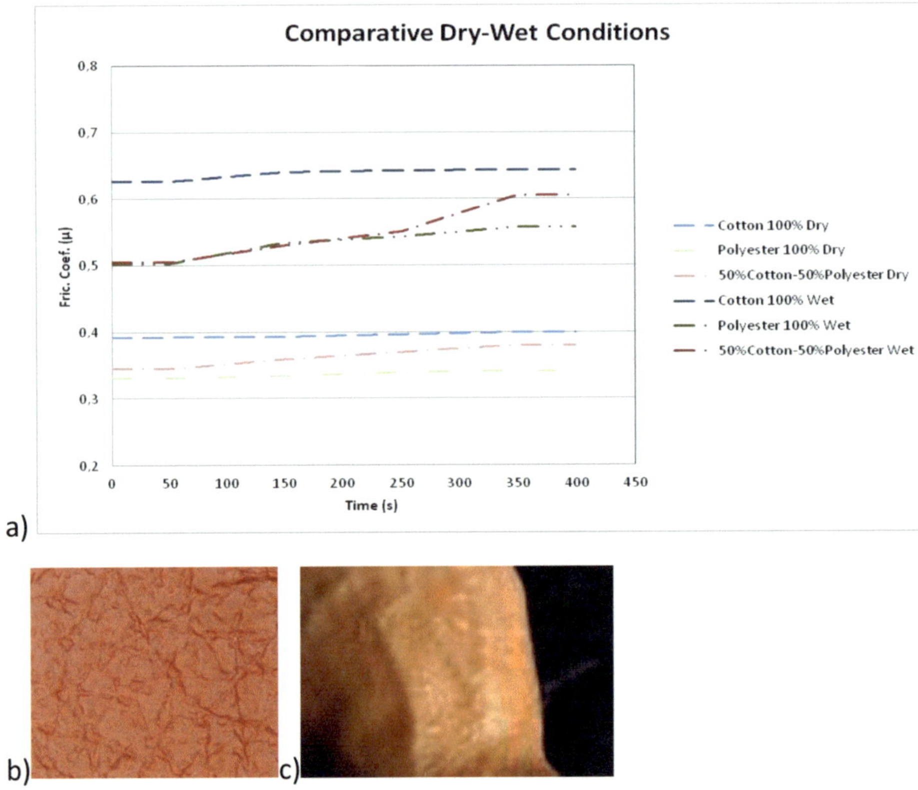

Figure 25.
(a) Friction measurements as indicator of textile tactility in dry and wet conditions for cotton, polyester, and the 50% mixture. (b) Material used to simulate the skin at laboratory. (c) Real skin.

5.5 Biotribology

5.5.1 Tribology of autologous bones

The use of impact and compression tests to understand the effect of thermal treatments in artificial bones and inserts has been studied in the frame of the EU Project AUTOBONE. In **Figure 26**, it can be appreciated that the thermal treatment reduces the wear caused by the impact, improving the toughness of the artificial inserts composed of hydroxyapatite and collagen [57].

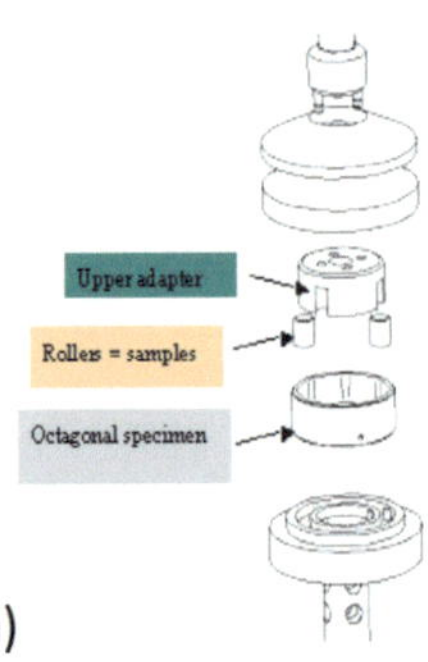

b)

Material	Speed (rpm)	Time (min)	Impacts per roller	Wear (mg)	Observations
HA/col	600	30	144,000	5.1 & 2.3; 6.3 & 5.5 **(4.8)**	Good impact wear resistance
HA/col CL	600	30	144,000	4.9 & 7.2; 4.1 & 5.1 **(5.3)**	Good impact wear resistance
HA/col/TT	600	30	144,000	2.0 & 1.8; 1.2 & 1.7 **(1.7)**	Very good impact wear resistance
A/col CL/TT	600	30	144,000	1.4 & 1.0; 2.0 & 2.6 **(1.8)**	Very good impact wear resistance

Figure 26.
(a) Configuration of impact test. (b) Impact resistance of the hydroxyapatite/collagen bone inserts with and without thermal treatments.

5.5.2 Tribology of hip and knee implants

A testing configuration has been designed adapted to a commercial FALEX tribometer to simulate the wear of hip/acetabular cup. Friction, wear, and tribocorrosion protocols have also been developed to simulate the lifetime of the hip and knee prosthesis. In **Figure 27**, it is possible to see how the TiCN-2 coating deposited by Physical vapor deposition (PVD) by TEKNIKER can enhance the tribocorrosion properties and antibacterial properties. The antibacterial properties are still enhanced when adding a sacrificial silver top layer, also applied by PVD. The Ag layer might be active as anti-infection layer, during implantation, since it will be easily removed due to their poor wear resistance. The protocols have been used in the national projects FUNCOAT, DELECA, SINOVIAL, FUNMAT in the industrial BIOTIDE project, and the Eurostars INNOVATIDE Project [29–31].

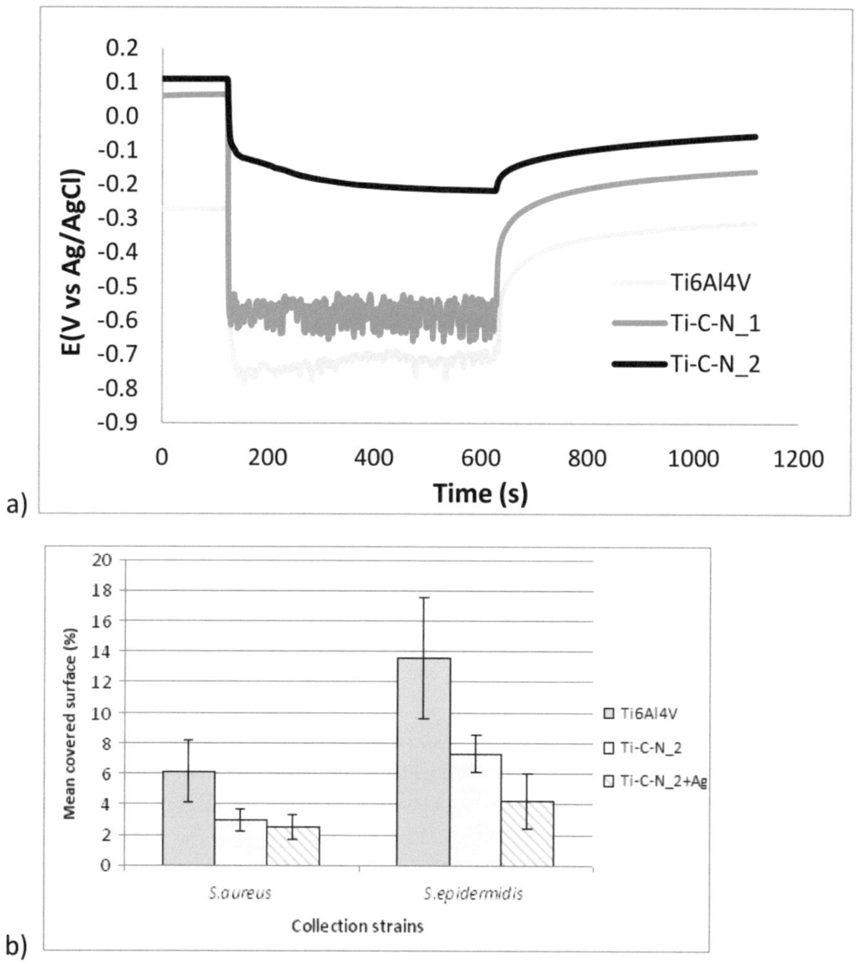

Figure 27.
(a) Program for coating evaluation. (b) Tribocorrosion resistance of Ti_6Al_4V alloy before and after coating with TiCN-1 and TiCN-2. (c) Antibacterial properties of Ti_6Al_4V, after coating with TiCN-2 without and with a top silver layer.

5.5.3 Dental tribology

The development of testing protocols can be highlighted to simulate friction, wear, impact, and tribocorrosion in dental implants. In **Figure 28**, the results of tribocorrosion tests carried out to compare the behavior of titanium before and after treatment with different plasma electrooxidation coatings (PEO) developed by TEKNIKER using different process conditions are represented. The behavior of the coatings improved when increasing the plasma intensity. Also, the wear volume of a reference and new electrolyte was compared observing a significative improvement with the new electrolyte developed in TEKNIKER. This knowledge has been applied to the national project FUNCOAT and industrial projects [58].

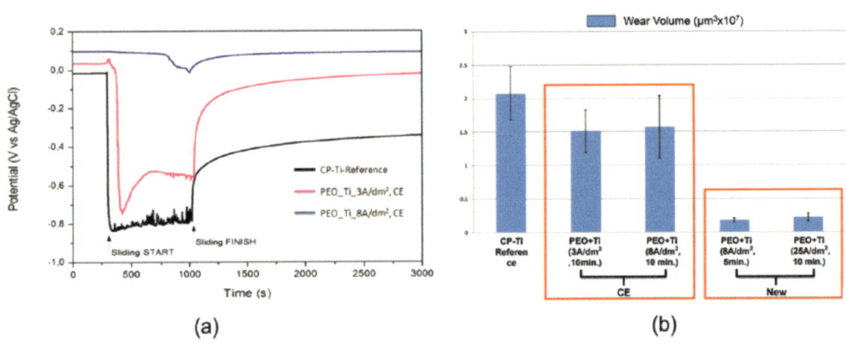

Figure 28.
(a) Tribocorrosion test of titanium (Ti) coated with two types of PEO coatings. (b) Wear volume of tribocorrosion test of Ti and PEO coatings using two electrolytes.

5.5.4 Tactile properties of steel

A similar test protocol mentioned in Section 5.4.3 has also been used to evaluate the steel sheet tactility. The effect of the textured surface geometries of steel sheet surfaces transferred by lamination with laser textured rollers has been studied. The laser texturing was carried out by TEKNIKER. This knowledge has been used in the EU Project STEELTAC [59].

5.6 Tribolubrication

5.6.1 Tribology of environmentally friendly lubricants for cutting and forming applications

Testing protocols has been developed to compare tapping torque results and center of numeric control (CNC) machines to screen torque and wear of cutting and forming fluids and different coatings to increase the efficiency. In the EU

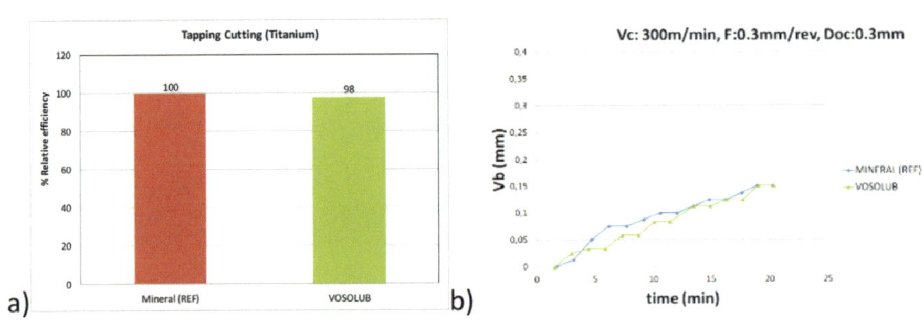

Figure 29.
Comparison of mineral and vegetable lubricants measuring (a) relative efficiency in tapping torque machine and (b) flank wear in CNC grinding machine.

Project VOSOLUB [60], similar tribological properties have been shown between mineral- and vegetable-based cutting lubricants. Both the efficiency of the fluids measuring the torque with the tapping torque machine and the flank wear with a CNC controlled grinding machine have been evaluated (**Figure 29**). The knowledge has been applied to EU Projects IBIOLAB [60], DECOLUB [61], and other related projects such as ECOLUBRO, NANOCOMP, MEFOLUB, and CLAREFOSS.

5.6.2 Tribology of biolubricants for bearings and gears

Testing protocols have been developed to compare the behavior of the biolubricants as alternative to mineral-based oils to lubricate mechanical components (e.g., wind mills, hydraulic pumps, excavators). In the example, a glycol-free lubricant has been developed increasing their load-carrying capacity in comparison with the glycol-containing one. The glycol-free lubricant is biodegradable, and it can be an interesting alternative as a fire-resistant hydraulic oil (**Figure 30**). The protocols have been used in the EU Projects VOSOLUB [38], SUNOIL and BIOGREASE, [39], LUBRICOAT [40], CTVO-NET and LLINCWA [62], BIOMON and IBIOLAB [41], the national projects LUVE and BIOVESIN [63, 64], and the industrial projects ADIVINA, PAVIA, and NANOINTECH.

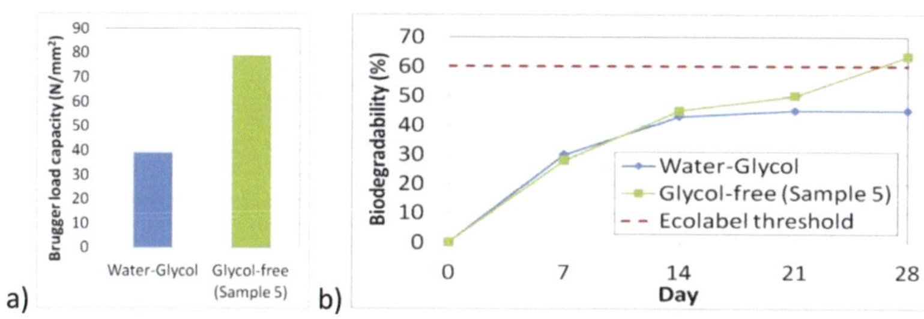

Figure 30.
(a) Load-carrying capacity of glycol and glycol-free lubricants. (b) Biodegradability of glycol-free lubricants >60%.

5.6.3 Tribology of ionic liquids

Testing protocols has been developed also to evaluate the behavior of ionic liquids as lubricants in different environments studying friction, wear, and tribocorrosion behavior. In **Figure 31**, it is represented the friction and wear scar of the Ionic Liquids DIL3 and

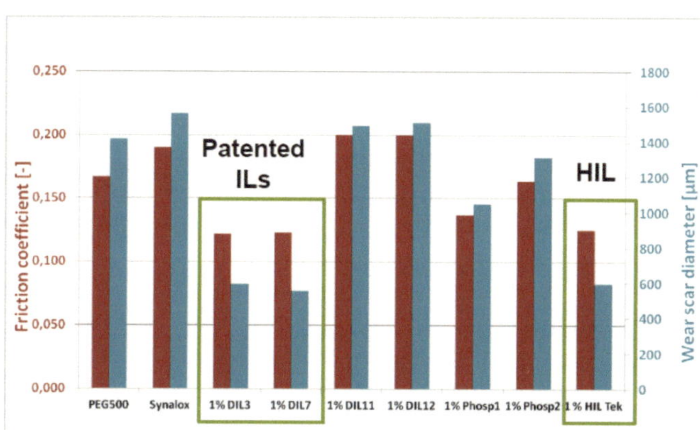

Figure 31.
In the figure, the friction and wear scar of the ionic liquids DIL3 and DIL7, patented by TEKNIKER, are compared with the hybrid ionic liquid synthetized by TEKNIKER.

DIL7, patented by TEKNIKER (EFS/ID/12720945) and developed in the frame of EU Project MINILUBES Project has been compared with a hybrid ionic liquid synthetized by TEKNIKER, developed in the EMAITEK project, are represented. In this subject, the activity carried out in Eurostar Project VACUUM DLC, the EMAITEK Basque Country Initiative, and the Austrian COMET Project can be also mentioned [65–67].

5.7 Tribology of materials for energy

5.7.1 Tribology for biogas, biodiesel, and bioethanol

Testing protocols has been set up to determine the durability of fuel injectors and other critical engine components of engines working with biogas or suffering lubricant dilution with biodiesel, bioethanol, or their mixtures. In **Figure 32**, it is represented the friction coefficient of the PVD coating Ti-DLC coating deposited by TEKNIKER on injectors, comparing the behavior between diesel and biodiesel (B50, 50% diesel and 50% biodiesel). The protocols have been used in the EU Project Cleanengine and in the industrial project IDEA [68].

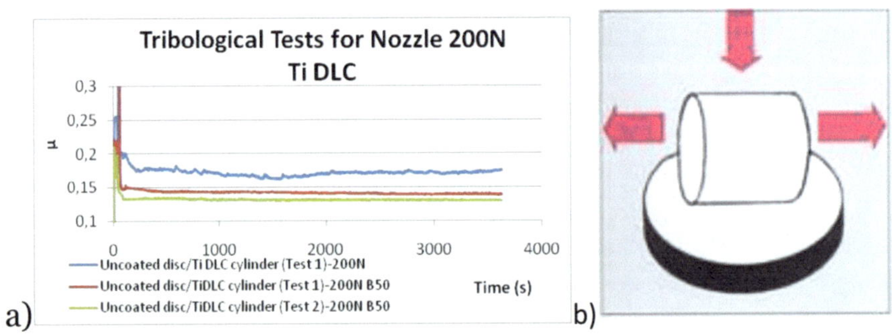

Figure 32.
(a) Tribological properties of the Ti-DLC coating, simulating the contact of injector nozzles with diesel and biodiesel B50; and (b) testing configuration.

5.7.2 Tribology in energy-efficient systems

Testing protocols have been set up to compare the lifetime of critical components of heat exchangers, compressors, gensets, and microturbines. In the example,

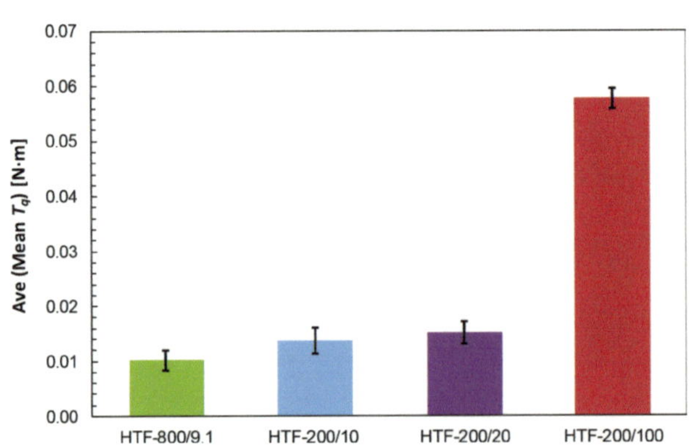

Figure 33.
Average main torque obtained with the drag friction test with different heat exchange fluids of different viscosities (mPa). Error bars represent average standard deviation.

the drag friction test developed by TEKNIKER (**Figure 3**), has been used to evaluate the average torque of different heat exchanger fluids, named as HTF-800 and HTF-200. The last one was tested at different viscosities from 10 to 100 mPa s. The minimum torque was measured for the HT-800 fluid, being finally selected for the application in SUSPIRE Project (**Figure 33**). The protocols have been applied to EU Project HEGEL [68] and to the industrial project ABADIE PLUS [69].

6. Conclusions

Attention to tribological problems would imply worldwide annual savings of 970,000 M€ (1.39% GDP) [69]. A huge range of new testing equipment and protocols have been developed to simulate the friction and wear (mechanical and chemical) of a wide range of mechanical components and systems, in different environments. The knowledge about the working conditions and the failure mechanisms will make possible in the future to simulate at laboratory a still wider list of tribological systems. This will also help us to make a step forward, toward modeling the extrapolation of research results to real systems and the generation of new standards. Artificial intelligence can be also a tool that can facilitate this step forward. Tribology will help to design low carbon footprint materials and implement them in a cost-efficient, predictive, and safe way [70–75]. Tribology will assist also the development of energy- and resource-efficient products and processes, in real systems.

Acknowledgements

The authors acknowledge the financing from the European Commission, the Spanish Minister of Science and Technology and CDTI, the Basque Country Government SPRI and InnoBasque, the Gipuzkoa Provincial Council and the X-Tribology Austrian COMET Program. The authors also acknowledge the cooperation and/or financing of the industries and research institutions participating in each project. The authors made special mention to those members of the tribology team, that completed their Doctoral Thesis for their contribution to the knowledge generated: Bihotz Pinedo (Tribology of seals 2016) [23-25, 27], Borja Zabala (high Temperature Tribology (eg. moulds, engines) 2017) [20, 22, 42, 43, 46, 52], Francesco Pagano, (tribology of ionic liquids and vacuum tribology, 2017) [2–4, 13, 57–65], Beatriz Fernandez-Diaz (tribology of polymers, 2017) [28, 34, 63], Virginia Saenz de Viteri (biotribology, 2016) [30, 31, 56, 57], Sofia Alves (dental tribology, 2017) [58], Cristina Cerrillo (ecotoxicity of nanoparticles, 2016) [71–75], Ainara López (marine and off shore tribology, 2018) [12–16]. Also, to all the members of the tribology team that dedicate their professional work to tribology activity Elena Fuentes [17, 18, 30, 40, 56, 57], Xana Fernandez [9, 11, 29, 31, 46–48, 54, 65], Gemma Mendoza [5, 11, 32, 49, 50, 60, 63, 71–73], Xanti Almandoz [34], Juan Carlos Rodriguez [8, 9, 16, 34], Olatz Areitioaurtena [48, 67, 68, 71–73], Marcello Conte [1, 2, 23–25, 27, 54], and Roman Nevshupa [1, 4, 7, 8, 46, 52, 54], and the other members of the research group that contributed to the generation of the results.

Author details

Amaya Igartua[1*], Raquel Bayon[1], Ana Aranzabe[1] and Javier Laucirica[2]

1 Fundación TEKNIKER, Eibar, Spain

2 IK4, Eibar, Spain

*Address all correspondence to: amaya.igartua@tekniker.es

A summary of 37 years' experience in TEKNIKER.

References

[1] Nevshupa A, Conte M, Igartua A, Roman E, Segovia JL. Ultrahigh vacuum system for advanced tribology studies: Design principles and applications. Tribology International. 2015;**86**:28-36. DOI: 10.1016/j.triboint.2015.01.020

[2] Totolin V, Conte M, Berriozabal E, Pagano F, Minami I, Dörr N, et al. Tribological investigation of ionic liquids in ultra-high vacuum environment. Lubrication Science. 2014;**26**:514-524. DOI: 10.1002/ls.1224

[3] Igartua A, Pagano F, Bayon R, Minami I, Doerr N. Ionic liquids in space applications. In: TAE Congress of Tribology; January 2018; Esslingen, Germany. 2018

[4] Igartua A, Berriozabal E, Nevshupa R, Roman E, Pagano F, Pleth Nielsen L, et al. Screening of diamond-like carbon coatings in search of a prospective solid lubricant suitable for both atmosphere and high vacuum applications. Tribology International. 2017;**114**:192-200. DOI: 10.1016/j.triboint.2017.04.024

[5] Silva ER, Ferreira O, Ramalho PA, Azevedo NF, Bayon R, Igartua A, et al. Ecofriendly non biocide reléase coatings for marine biofouling prevention. Science of the Total Environment. 2019;**650**(2):2499-2511. DOI: 10.1016/j.scitotenv.2018.10.010

[6] Tulcidas AV, Bayon R, Igartua A, Bordado JMC, Silva ER. Friction reduction on recent non releasing biocidal coatings by a newly designed friction test rig. Tribology International. 2015;**91**:140-150. DOI: 10.1016/j.triboint.2015.06.017

[7] Bayón R, Nevshupa R, Zubizarreta C, Ruiz de Gopegui U, Barriga J, Igartua A. Characterisation of tribocorrosion behaviour of multilayer PVD coatings. Analytical and Bioanalytical Chemistry. 2010;**396**(8):2855-2862. DOI: 10.1007/s00216-010-3538-6

[8] Bayon R, Zubizarreta C, Nevshupa R, Rodríguez JC, Fernández X, Ruiz de Gurpegui U, et al. Rolling-sliding, scuffing and tribocorrosion behaviour of PVD multilayer coatings for gear applications. Tribology International. 2011;**63**(1):17-26. DOI: 10.1108/00368791111101803

[9] Bayon R, Igartua A, Fernandez X, Martinez R, Rodriguez JC, Garcia JA, et al. Corrosion-wear behaviour of PVD Cr/CrN multilayer coatings for gear applications. Tribology International. 2009;**42**:591-599. DOI: 10.1016/j.triboint.2008.06.015

[10] Celis JP, Ponthiaux P, Bayón R, Igartua A. Testing Tribocorrosion of Passivating Materials Supporting Research and Industrial Innovation: Handbook. London: Maney Publishing; 2012. European Federation of Corrosion. Available from: https://www.taylorfrancis.com/books/e/9781351546942

[11] Igartua A, Bayon R, Mendoza G, Fernandez X, Ciarsolo I, Aramburu E, et al. Corrosion and tribocorrosion studies on PVD coatings for aeronautical applications. In: Workshop on Surface Treatments and Coatings for Mechanical and Aeronautical Applications; 28-30 Marzo 2007; Sevilla. 2007

[12] López-Ortega A, Arana JL, Bayón R. Tribocorrosion of passive materials: A review on test procedures and standards. International Journal of Corrosion. 2018;**1**:7345346. DOI: 10.1155/2018/7345346

[13] López A, Bayón R, Pagano F, Igartua A, Arredondo A, Arana JL, et al. Tribocorrosion behaviour of mooring high strength low alloy steels in synthetic seawater. Wear.

2015;**338-339**:1-10. DOI: 10.1016/j.wear.2015.05.004

[14] López-Ortega A, Bayón R, Arana JL, Arredondo A, Igartua A. Influence of temperature on the corrosion and tribocorrosion behaviour of high strength low alloy steels used in offshore applications. Tribology International. 2018;**121**:341-352. DOI: 10.1016/j.wear.2015.05.00

[15] López-Ortega A, Bayón R, Arana JL. Evaluation of protective coatings for offshore applications. Corrosion and tribocorrosion behavior in synthetic seawater. Surface & Coatings Technology. 2018;**349**:1083-1097. DOI: 10.1016/j.surfcoat.2018.06.089

[16] López-Ortega A, Arana JL, Rodríguez E, Bayón R. Corrosion, wear and tribocorrosion performance of a thermally sprayed aluminum coating modified by plasma electrolytic oxidation for offshore submerged components protection. Corrosion Science. 2018;**143**:258-280. DOI: 10.1016/j.corsci.2018.08.001

[17] Fuentes E, Kanase K. Friction and wear in engine bearings. In: Critical Component Wear in Heavy Duty Engines. Singapore: John Wiley & Sons; 2011. DOI: 10.1002/978047082 8847.ch12. ISBN 978-0-470-82882-3

[18] Kallio M, Vuorinen P, Fuentes E, Maraña O, Ruusila V, Nyysönnen T, et al. Tribological behaviour of bronze alloys with solid lubricants. Key Engineering Materials. 2013;**527**:205-210

[19] Igartua A, Laucirica J, Aranzabe A. Application of low temperature PVD coatings in rolling bearings. Tribological tests and experiences with spindle bearing systems. Surface and Coating Technology. 1996;**1**:460-467. DOI: 10.1016/ S0257-8972(96)03050-2

[20] Zabala B. Doctoral thesis. Universidad País Vasco; 2017

[21] Gorjan L, Boretius M, Guglan G, Kuebler J, Igartua A. Ceramic protection plates brazed to aluminum brake discs. Ceramics International. 2016;**42**(14):15739-15746. DOI: 10.1016/j.ceramint.2016.07.035

[22] Zarraga O, Abete JM, Ulacia I, Zabala B, Uzkudun O. On the Development of a Simple Model of a Brake Clutch for Squeal Prediction, Congress Eurobrake. 2014. pp. 13-14

[23] Pinedo B, Aguirrebeitia J, Conte M, Igartua A. Tri-dimensional eccentricity model of a rod lip seal. Tribology International. 2014;**78**:68-74. DOI: 10.1016/j.triboint.2014.05.005

[24] Conte M, Pinedo B, Igartua A. Frictional heating calculation based on tailored experimental measurements. Tribology International. 2014;**74**:1-6. DOI: 10.1016/j.triboint.2014.01.020

[25] Conte M, Pinedo B, Igartua A. Role of crystallinity on wear behaviour of PTFE composites. Wear. 2013;**307**: 81-86. DOI: 10.1016/j.wear.2013.08.019

[26] Lin LY, Tlatlik H, Gralla R, Igartua A, Baets P, Schlarb AK. Mechanical and thermal behaviors of polyether etherketone-based multi-scale composites. Journal of Composite Materials. 2012;**2**(10):1-18. DOI: 10.1177/0021998312454317

[27] Pinedo B, Hadfield M, Tzanakis I, Conte M, Anand M. Thermal analysis and tribological investigation on TPU and NBR elastomers applied to sealing applications. Tribology International. 2018;**127**:24-36. DOI: 10.1016/j.triboint.2018.05.032

[28] Fernandez B. Tesis doctoral; Universidad del País Vasco; 2017

[29] Saenz de Viteri V, Barandika G, Bayon R, Fernandez X, Ciarsolo I, Igartua A, et al. Development of Ti-C-N coatings with improved

tribological behavior and antibacterial properties. Journal of the Mechanical Behavior of Biomedical Materials. 2016;**55**:75-86. DOI: 10.1016/j.jmbbm.2015.10.020

[30] Sáenz de Viteri V, Fuentes E. Titanium and Titanium Alloys as Biomaterials. Rijeka, Croatia: InTech. DOI: 10.5772/55860

[31] Sáenz de Viteri V, Bayon R, Igartua A, Barandika G, Moreno JE, Pérez C, et al. Structure, tribocorrosion and biocide characterization of Ca, P and I containing TiO_2 coatings developed by plasma electrolytic oxidation. Applied Surface Science. 2016;**367**:1-10. DOI: 10.1016/j.apsusc.2016.01.145

[32] Igartua A, Bayon R, Mendoza G. Plasma electrolytic oxidation coatings for engine components. Ricerca e Sviluppo. 2014;**67**:44-52

[33] Laucirica J, Arnaiz A, Igartua A. Mantenimiento y control mediante Sistema experto; Forum Calidad. Vol. 361992. pp. 54-61

[34] Presa J, Fernández-Díaz B, Igartua A, San José E, Almandoz X, Rodríguez JC, et al. Effects of manufacturing errors in the behaviour of a power gearbox. In: LUBMAT Conference; June 2016. 2016. Available from: https://www.slideshare.net/TogetherToSolve/effects-of-manufacturing-errors-in-the-behaviour-of-a-power-gearbox

[35] Martins R, Seabra J, Brito A, Seyfert C, Luther R, Igartua A. Friction coefficient in FZG gears lubricated with industrial gear oils: Biodegradable ester vs. mineral oil. Tribology International. 2006;**39**:512-521. DOI: 10.1016/j.triboint.2005.03.021

[36] Cardoso NFR, Martins RC, Seabra JHO, Igartua A, Rodriguez JC, Luther R. Micropitting performance of nitrided steel gears lubricated with mineral and ester oils. Tribology International. 2008;**42**(1):77-87. DOI: 10.1016/j.triboint. 2008.05.010

[37] Martins RC, Cardoso NFR, Bock H, Igartua A, Seabra JHO. Power loss performance of high pressure nitrided steel gears. Tribology International. 2009;**42**:1807-1815. DOI: 10.1016/j.triboint.2009.03.006

[38] Leao JD, Bouillon V, Muntada L, Johnson C, Wilson P, Vergnes O, et al. New formulations of sunflower based bio-lubricants with high oleic acid content. Oilseeds and fats, Crops and Lipids. 2016;**23**(5):D509. DOI: 10.1051/ocl/2016033

[39] Igartua A, Barriga J, Aranzabe A. Biodegradable Lubricants. Vol. 1. Radom, Poland: Virtual Tribological Institute; 2005. pp. I.1-VI.3. ISBN 83-7204-449-X

[40] Vercammen K, Van Acker K, Barriga J, Fuentes E, Kalin M, Vizintin J. Combining biolubricants and low friction coatings. In: Current Trends in Tribology. Vol. I. Poland: ITEE; 2005. pp. 1-4. ISBN 83-70204-418-X

[41] Igartua A. Biolubricants & their control for sustainable developments. Lube. 2007;**81**:6. ISSN 1744-5418

[42] Zabala B, Igartua A, Scarpi V, Timelli G, Girot F. Multiparametric study of Leidenfrost point and wettability of lubricants on high-pressure die-casting dies. International Journal of Thermal Sciences. 2018;**125**:66-73. DOI: 10.1016/j.ijthermalsci.2017.11.014

[43] Zabala B, Igartua A, Raone C, Timelli G, Bonollo F, Scarpi V, Peggiato M, Girot F, "Evaluation HPDC Lubricant Spraying for Improved Cooling and Die Protection, LUBMAT Conference, 2016. Available from: https://www.researchgate.net/publication/315837772_Evaluation_HPDC_Lubricant_Spraying_for_Improved_Cooling_and_Die_Protection

[44] Barriga J, Igartua A, Marín M. Mejora del proceso de forja mediante la aplicación de nitruración a alta presión y la introducción de una lubricación ecológica. Boletin de la Sociedad Espanola de Ceramica y Vidrio. 2004;**13**(2). DOI: 10.3989/cyv.2004.v43.i2.491

[45] Caliari D, Timelli G, Igartua A. Microstructural and tribological investigations of diecast and hard anodized AlSiCu alloys. Surface and Coatings Technology. 2018;**352**:462-273. DOI: 10.1016/j.surfcoat.2018.07.084

[46] Zabala B, Igartua A, Fernandez X, Nevshupa R. Friction and wear of a piston ring/cylinder liner at the top dead center: Experimental study and modelling. Tribology International. 2017;**106**:23-33. DOI: 10.1016/j.triboint.2016.10.005

[47] Igartua A, Fernandez X, Woydt M, Luther R, Illarramendi I. Tribological tests to simulate wear in piston rings. In: Critical Component Wear in Heavy Duty Engines. Vol. 1. 2011. pp. 167-195. DOI: 10.1002/9780470828847.ch11

[48] Igartua A, Fernández X, Areitioaurtena O, Luther R, Seyfert C, Rausch J, et al. Biolubricants and triboreactive materials for automotive applications. Tribology International. 2009;**42**:561-568. DOI: 10.1016/j.triboint.2008.10.015

[49] Picas JA, Forn A, Igartua A, Mendoza G. Mechanical and tribological properties of high velocity oxy-fuel thermal sprayed nanocrystalline CrC-NiCr coatings. Surface and Coating Technology. 2013;**174-175**:1095-1100. DOI: 10.1016/S0257-8972(03)00393-1

[50] Bellido-González V, Igartua A, Mendoza G, Powell S, Talks MG, Momeñe E. Replacement of hard-chrome in cylinder liners. A PVD approach. Society of Vacuum Coaters. 2004:505-856. ISBN 0737-5921

[51] Igartua A, Tonder K. COST 532-Triboscience and tribotechnology superior friction and wear control in engines and transmissions. In: Engine Systems. Belgium: COST Edition; 2008. pp. 108-113. ISBN: 978-92-898-0040-2

[52] Fdez-Pérez X, Igartua A, Nevshupa R, Zabala P, Zabala B, Luther R, et al. Innovative "green" tribological solutions for clean small engines. In: Gegner J, editor. Tribology-Fundamentals and Advancements. Rijeka, Croatia: IntechOpen; DOI: 10.5772/55836. Available from: https://www.intechopen.com/books/tribology-fundamentals-and-advancements/innovative-green-tribological-solutions-for-clean-small-engines

[53] Gili F, Igartua A, Luther R, Woydt M. The impact of biofuels on engine oil performance. Lubrication Science. 2011;**23**(7):313-330. DOI: 10.1002/ls.158

[54] Igartua A, Nevshupa R, Fernandez X, Conte M, Zabala R, Zabala P, et al. Alternative eco-friendly lubes for clean two-stroke engines. Tribology International. 2011;**44**:727-736. DOI: 10.1016/j.triboint.2010.01.019

[55] Available from: http://slipsafe.org/

[56] Igartua A, Saenz de Viteri V, Fuentes E. Tribological response of the cotton textile against human skin equivalents. In: The International Istambul Textile Congress. 30th May-1st June 2013

[57] Fuentes E, Sáenz de Viteri V, Igartua A, Martinetti R, Dolcini L, Barandika G. Structural characterization and mechanical performance of Ca/P scaffolds and natural bones: A comparative study. Journal of Applied Biomaterials & Biomechanics. 2010;**8**:159-165. DOI: 10.5301/JABB.2010.6092

[58] Alves SA, Bayon R, Igartua A, Saenz de Viteri V, Rocha A. Tribocorrosion behavior of anodic titanium oxide

films produced by plasma electrolytic oxidation for dental implants. Lubrication Science. 2014;**26**:500-513. DOI: 10.1002/ls.1234

[59] (a) Zhang S, Zeng X, Matthews DTA, Igartua A, Rodriguez-Vidal E, Contreras Fortes J, et al. Selection of micro-fabrication techniques on stainless steel sheet for skin friction. Friction. 2016;**4**:89-104. DOI: 10.1007/s40544-016-0115-9. (b) Rodriguez E, Igartua A. Texture design for reducing tactile friction independent of sliding orientation on stainless steel sheet. Tribology Letters. 2017;**3**:65-89. DOI: 10.1007/s11249-017-0869-x

[60] Mendoza G, Igartua A, Arizaga I, Azpiazu F, Santiso G. Metal working fluids based on vegetable oils for grinding operations. Journal of Mechanics Engineering and Automation. 2012;**2**(10):638-646

[61] Igartua A, Barriga J, Pérez J, Muntada L, Navío P. Ecological lubricants for cutting and forming applications. In: New Horizons for Tribology and Lubricants; Expert Verlag 2002. Vol. 10. Austria: OETG; 2001. pp. 60-64. Available from: http://www.oetg.at/fileadmin/Dokumente/oetg/Proceedings/WTC_2001_files/html/M-29-13-751-IGARTUA.pdf

[62] Igartua A. Biodegradable lubricants. Lube. 2007;**82**:19-22. ISSN 1744-5418

[63] Igartua A, Mendoza G, Fernandez-Diaz B, Urquiola F, Vivanco S, Arguizoniz R. Vegetable oils as hydraulic fluids for agricultural applications. Grasas y Aceites. 2011;**62**:29-38. DOI: 10.3989/gya.056210. ISBN: 0017-3495

[64] Igartua A Synthetic lubrication and tribotest. Lubrication Science. 2014;**26**:447-448. DOI: 10.1002/ls.1277

[65] Pagano F, Zare P, Mahrova M, Dörr N, Bayon R, Fernandez X, et al. Dicationic ionic liquids as lubricants. Journal of Engineering Tribology. 2012;**117**:359-366

[66] Kronberger M, Pagano F, Pejakovic V, Igartua A, Urbistondo E, Kalin M. Miscibility and tribological investigation of ionic liquids in biodegradable esters. Lubrication Science. 2014;**26**:463-387. DOI: 10.1002/ls.1274

[67] Stolte S, Steudte S, Areitioaurtena O, Pagano F, Thöming J, Stepnowski P, et al. Ionic liquids as lubricants or lubrication additives: An ecotoxicity and biodegradability assessment. Chemosphere. 2012;**89**:1135-1141. DOI: 10.1016/j.chemosphere.2012.05.102

[68] Gilabert E, Gorritxategi E, Conde E, Garcia A, Areitioaurtena O, Igartua A. An advanced maintenance system for polygeneration applications. In: Engineering Asset Lifecycle Management. 2010. pp. 823-830. Available from: https://rd.springer.com/chapter/10.1007%2F978-0-85729-320-6_94

[69] Holmberg K, Erdermir A. Influence of tribology on global energy consumption, costs and emissions. Friction. 2017;**5**(3):263-284. DOI: 10.1007/ s40544-017-0183-5

[70] Igartua A, Urdangarín L, Pérez de Larraya D. Motores de barco adaptados al uso de lubricantes y fueles alternativos, Europa Azul. 2008. pp. 68-69

[71] Cerrillo C, Barandika G, Igartua A, Areitioaurtena O, Mendoza G. Key challenges for nanotechnology: Standardization of ecotoxicity testing. Journal of Environmental Science and Health, Part C. 2016;**4**:89-104. https://doi.org/10.1002/etc.2999

[72] Cerrillo C, Barandika G, Igartua A, Areitioaurtena A, Uranga N, Mendoza G. Colloidal stability and ecotoxicity of multiwalled carbon

nanotubes: influence of select organic matters. Environmental Toxicology and Chemistry. 2016;**35**:74-83. https://doi.org/10.1002/etc.3172

[73] Cerrillo C, Barandika G, Igartua A, Areitioaurtena O, Mendoza G. Towards the standardization of nanoecotoxicity testing: Natural organic matter 'camouflages' the adverse effects of TiO_2 and CeO_2 nanoparticles on green microalgae. Science of the Total Environment. 2016;**543**(A1):95-104. https://doi.org/10.1016/j.scitotenv.2015.10.137

[74] Laux P, Riebeling C, Booth AM, Brain JD, Brunner J, Cerrillo C. Challenges in characterizing the environmental fate and effects of carbon nanotubes and inorganic nanomaterials in aquatic systems. Environmental Science: Nano. 2018;**5**:48-63. DOI: 10.1039/C7EN00594F

[75] Laux P, Riebeling C, Booth AM, Brain JD, Brunner J, Cerrillo. Biokinetics of Nanomaterials: the Role of Biopersistence. NanoImpact. 2017;**6**:69-80. DOI: 10.1016/j.impact.2017.03.003

Chapter 10

High-Temperature Self-Lubricating Metal Nitride-Based Nanostructure Composite Films

Hongbo Ju

Abstract

The film technology is one of the most efficient methods of modifying the surface properties of materials. The focus of recent attention in the solid lubricating materials is on the hard transition metal nitride films' excellent high-temperature tribological properties applied in powertrains and cutting tools. This chapter reviewed the formation mechanism of Magnéli phases and the influence of Magnéli phases on the high-temperature tribological properties of hard transition metal nitride films synthesized using magnetron sputtering. The self-lubricating behavior and wear mechanism of the Magnéli phases are discussed from the point of view of crystallography. Some methods to enhance the wear resistance property of the metal nitride film are also discussed.

Keywords: transition metal nitride films, magnetron sputtering, tribological properties, self-lubricating, Magnéli phase

1. Introduction

Recently, a solid lubrication known as Magnéli phases was reported in the application of oxide materials with "easy" crystallographic shear phases, and numerous related research activities have been carried out to investigate the function mechanism of Magnéli phases. Molybdenum, tungsten, and vanadium as the additional elements were initially incorporated into transition metal nitride (TMN)-based hard films, since the molybdenum oxide, tungsten oxide, and vanadium oxide are common Magnéli phases and exhibit the excellent lubricating property. The TMN matrix as the hard phase with a columnar structure could retain the strength and bearing capacity during the wear test. Both the environment heat and friction heat induce the complex tribo-chemistry reaction and the formation of Magnéli phases. The binary molybdenum nitride, the scientists have already attached importance to tungsten nitride and vanadium nitride film in solid lubricating films in recent years. In this part, the high-temperature self-lubricating property of TMN-based hard films was discussed.

2. Binary molybdenum nitride, tungsten nitride, and vanadium nitride films

The crystal structure of the molybdenum nitride film shows a little influence on the deposition parameters such as nitrogen partial pressure, substrate temperature, and target power. A single fcc-Mo_2N phase is usually investigated in the molybdenum nitride film deposited using magnetron sputtering system.

Hardness of the molybdenum nitride film is ~26 GPa. The film exhibits the excellent friction property at high temperature at the expense of wear resistance property. Tribo-film MoO_3 is considered as the main factor attributing to the relatively low friction coefficient. MoO_3 is composed of double layers of distorted edge-sharing MoO_6 octahedra parallel to (010) planes. The MoO_3 has low shear strength because weak van der Waals forces hold the successive layers together. The counterpart is easy to wear away the MoO_3 under the wear test.

The crystal structure of the tungsten nitride film depends on the deposition parameters significantly. Tungsten nitride films could present a variety of phases such as cubic W_2N and hexagonal WN with large nitrogen to tungsten ratios, and the film consisting of a single phase of fcc-W_2N exhibits the highest hardness and lowest friction coefficient. The tungsten or nitrogen vacancies in the binary tungsten nitride films are a common phenomenon [1, 2]. The vacancies in the films could enhance the mechanical properties. The hardness of fcc-W_2N film is ~30 GPa, whose value is higher than that of molybdenum nitride and vanadium nitride films. Binary tungsten nitride film also exhibits the excellent friction property at the expense of wear resistance property. During the wear test, WO_3 could easily form induced by the counterpart at high temperature, and it could play an excellent lubricating role. WO_3 is described as crystallographic shear structure in normal as a type of typical Magnéli phases. Reeswinkel et al. [3] expounded the structure of WO_3 by the calculation results according to the density functional theory. The results show that there are three different types of W-O bonds co-exist in WO_3 owing to the distorted octahedron WO_6 and the shifted W cation in the WO_3. It is showed that these three types of W-O bonds represent extraordinary lubrication properties.

The crystal structure of vanadium nitride film is also influenced by the deposition parameters significantly. The vanadium nitride-based film is seldom applied in the cutting tools on account of the relative low value of hardness and poor thermal stability of the vanadium nitride film. Vanadium oxide phases could lubricate the film during the wear test at high temperature, and V_nO_{2n+1} was defined as Magnéli phase because of its excellent lubricant properties. The increase in temperature was reported to induce the change of VO_2 to V_2O_5 [4]. Erdemir [5] establishes the relationship between the phase lubricity of oxide and its ionization potential based on the principle of crystal chemistry. Based on this principle, oxides with high ionic potential values will show a low coefficient of friction. Because the ionic potential of V_2O_5 (10.2) is higher than that of VO_2 (6.8) [6], the V_2O_5 shows a better lubricant property than that of VO_2.

3. Molybdenum nitride, tungsten nitride, and vanadium nitride-based composite films

In view of the poor thermal stability and high wear rate, V-N, W-N, and Mo-N-based films are comparatively less applied in cutting tool field. As reported [7, 8], alloying metallic/nonmetallic elements into metal nitride matrix to deposit ternary, quaternary, and multicomponent hard films could combine the benefits of individual components. Therefore, the incorporation of some additional elements into the V-N, W-N, and Mo-N film could be an efficient method to improve the thermal stability and wear resistance property.

Addition of Al into transition metal nitride thin films has been regarded as an effective method to improve the hardness, crystalline orientation, wear resistance, oxidation resistance, and thermal stability of the materials [9, 10]. AlN alloys are usually considered as a superior oxidation resistant due to the formation of Al_2O_3 layers, which prevent oxygen diffusion toward the coating interior. In addition, Al and Al alloys exhibit low wear resistance although they are used in a wide range of automobile and aerospace industries [11]. Yang et al. [12] prepared Mo1-x Alx N thin films on stainless steel coins, alumina sheet, and silicon wafer substrates by using dc reactive magnetron sputtering technique and studied the effect of N2 and Al content (0.06–0.33) on the coating properties including structural, hardness, and oxidation resistance. The maximum hardness of 29 GPa was found at x = 0.06, and by further increasing of Al content, a decrease in hardness was detected owning to the weakness of fine grain strengthening. The oxidation resistance temperature increased gradually by increasing the concentration of Al. Our group [13] fabricated Mo-Al-N thin films with various Al contents (3.7–18.3 at.%) on stainless steel (06Cr19Ni10) and Si (100) wafer substrate by using reactive magnetron sputtering and studied the effect of Al content on microstructure, mechanical oxidation resistance, and tribological properties of the films. The result showed that the oxidation resistance of MoAlN thin films increased by increasing the Al content, while hardness and young modulus first increased and then started decreasing by increasing the Al content. The highest values of hardness and elastic modulus were 32.6 and 494 GPa, respectively, at 3.7 at.% Al. The film showed the lowest average friction coefficient and wear at the range between 4.1 and 9.5 at.% of Al content. Besides this, we also add the Al into the vanadium nitride film using the magnetron sputtering. The influence of Al content on the micro-structure, oxidation resistance, mechanical, and tribological properties of V-Al-N films were investigated. The crystal structure of V-Al-N film is always a single face-centered cubic structure no matter what the aluminum content is. When the content of aluminum was lower than 4.7 at.%, the hardness of the film represented an obvious increase. At the same time, the friction coefficient and wear rate showed a decrease at room temperature. The elevated Al content increases the oxidation resistance while reducing the fracture toughness. The high temperature tribological properties of the film at 4.7 at.% aluminum, which showed the highest hardness, lowest friction coefficient, and wear rate were investigated. The increase in temperature caused a change in the wear mechanism and a phase transition of the tribo-film vanadium oxides. The friction coefficient first increases to 0.7 at 300°C and decreases to 0.28 at 700°C with the wear rate gradually increases. As a result, the film at 4.7 at.% Al represented the most suitable properties in cutting tool application.

Based on above investigation, the addition of Al into the Mo_2N, VN matrix to the formation of the substitutional solid solution could enhance the mechanical properties. Besides this, the addition of Al also could improve the oxidation resistance temperature. The oxidation behavior of the films containing Al is mainly controlled by the Al content in the films. Al^{3+} diffuses outward to the surface of the films with the inward diffusion of oxygen will form Al_2O_3 layer during the oxidation process. Al_2O_3 phase is stable over a wide temperature range and easy to be formed. At low temperature ranging from 400 to 600°C, amorphous Al_2O_3 is formed [9]. The amorphous Al_2O_3 layer on the surface of the films is more effective to prevent the diffusion of oxygen into the films. However, the further increase in temperature induces the Al_2O_3 from amorphous to crystalline. Compared with the amorphous Al_2O_3, oxygen is easily diffused internally due to the grain boundary [10, 13, 14] of the crystalline oxide layer, while Al_2O_3 and MoO_3 grow rapidly. The oxidation diffusion mechanism is transferred from atomic diffuse to mass transfer. At the same time, the exchange of Al caused an increase in Al-N covalent bonds in the films, which represents excellent thermal stability property. The addition of Al could improve the wear resistance property of the films due to the enhancement of thermal stability and the decrease in the Magnéli elements in the films.

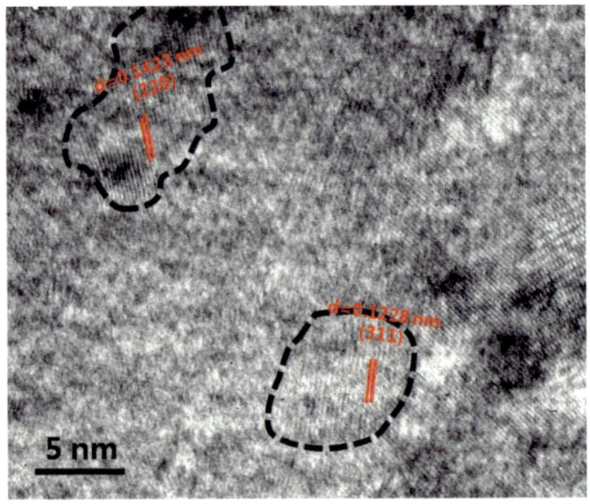

Figure 1.
HRTEM image of W-Si-N film with a Si content of 23.5 at %.

Silicon is another additional element to improve the mechanical, thermal stability, and wear resistance properties of the films. Limited solid solubility of Si in the TMN matrix induces the formation of amorphous Si_3N_4 phase. **Figure 1** shows the TEM image of the W-Si-N composite film with a Si content of 23.5 at.%. As shown in the figure, the nanograins of fcc-W_2N are embedded into the amorphous phases. The amorphous Si_3N_4 exhibits an excellent thermal stability property, and its oxidation resistance temperature is above 1200°C. Therefore, the addition of Si into the TMN matrix could increase the oxidation resistance temperature. The nanograins of tungsten nitride are wrapped up by the amorphous phase. This microstructure could provide better protection for tungsten nitride against oxidation and prolong the service life during the wear test.

Some other additional elements such as titanium and niobium also could improve the wear resistance properties of molybdenum nitride, tungsten nitride, and vanadium niride-based films [15–17]. For instance, the incorporation of titanium into the tungsten nitride matrix could form a solid solution of $(W_{1-x}Ti_x)_{2-y}Ny$ and exhibited a single face-centered cubic (fcc) W_2N structure. Both the sub-stoichiometric nitrogen content and solid solution strengthening could enhance the mechanical properties. The incorporation of titanium also drops the wear rate at room temperature significantly. This could be attributed to the increase of hardness to elastic modulus ratio, elastic recovery, and hardness. Besides this, the addition of titanium into the tungsten nitride matrix decreases the tungsten content in the film, and this finally induces the decrease in the content of Magnéli phase on the wear track surface. The decrease in the poor wear resistance Magnéli phase also attributes to the decline of wear rate. Tungsten titanium nitride film exhibits the higher working temperature than the binary W_2N film.

4. TMN-based composite films containing Magnéli elements

The molybdenum, tungsten, and vanadium Magnéli elements have been widely incorporated into some traditional TMN-based films to improve the high-temperature self-lubricating property. For instance, V was incorporated into the Nb-Si-N matrix, and the Nb-V-Si-N composite films were synthesized using the magnetron sputtering to improve the high-temperature tribological properties. The results show that the incorporation of V into the matrix could form the substitutional solid solution of (Nb, V)N. **Figure 2** shows the HRTEM image of the Nb-V-Si-N

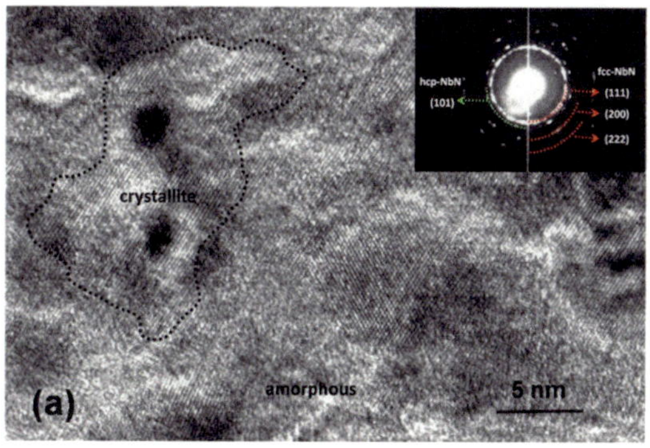

Figure 2.
HRTEM image and its corresponding SAED pattern of Nb-V-Si-N film at 3.7 at.% vanadium.

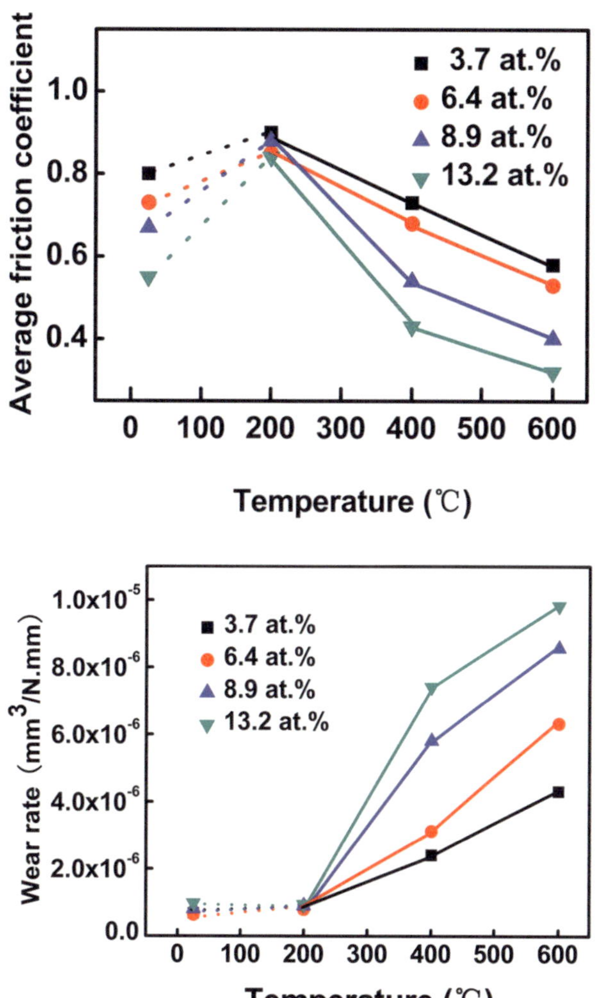

Figure 3.
Friction coefficient and wear rate of Nb-V-Si-N films as a function of vanadium content.

film with a vanadium content of 3.7 at.%. The amorphous Si_3N_4 phases enclose the substitutional solid solution. This microstructure could provide better protection for Nb-V-N against oxidation at the high temperature environment. The solution of V could improve the mechanical properties such as the hardness and toughness. **Figure 3** illustrates the friction coefficient and wear rate of the films at different testing temperatures. The incorporation of V into the film could improve the friction property at elevated temperatures at the expense of wear resistance property due to the Magnéli phase with weakly bonded lattice planes.

Molybdenum can also bring positive influence to high-temperature friction performance of the film. We add molybdenum into the classic titanium nitride film in order to improve its tribological properties. The result shows that the Ti-Mo-N film exhibits a single face-centered cubic structure, as same as binary TiN film. With the increase of molybdenum content, the Mo_2N phase appears in the film because of the precipitated molybdenum. **Figure 4** illustrates the SAED pattern of the Ti-Mo-N film at 46.0 at.% Mo. Calculated the data given in **Figure 4**, it can be seen that the selected electron diffraction pattern matches fcc-Mo_2N. The composite film represents the highest hardness and the lowest friction coefficient in room temperature at 46.0 at.% Mo. **Figure 5** shows the average friction coefficient and wear rate of Ti-Mo-N films at 46.0 at.% Mo under different testing temperatures. **Figure 6** shows the relative mass fraction of Ti-Mo-N, MoO_3, and TiO_2 on the wear tracks of Ti-Mo-N films at testing temperatures. When the texting temperature is higher than

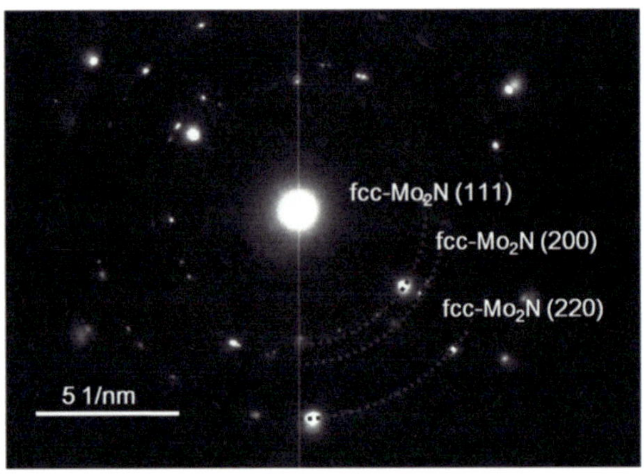

Figure 4.
The SAED patterns of Ti-Mo-N films with 46.0 at % Mo content.

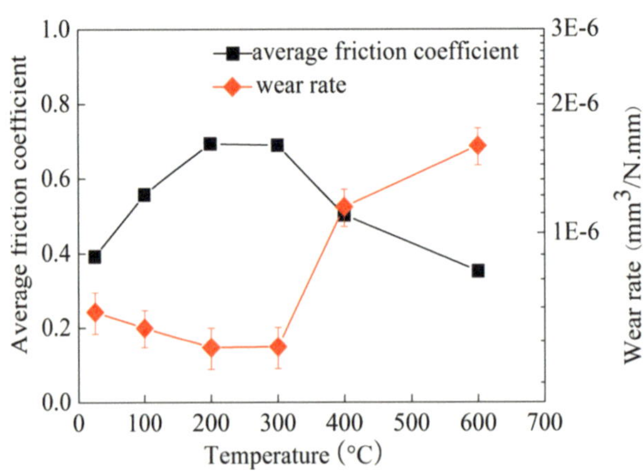

Figure 5.
The average friction coefficient and wear rate of Ti-Mo-N films at different temperatures.

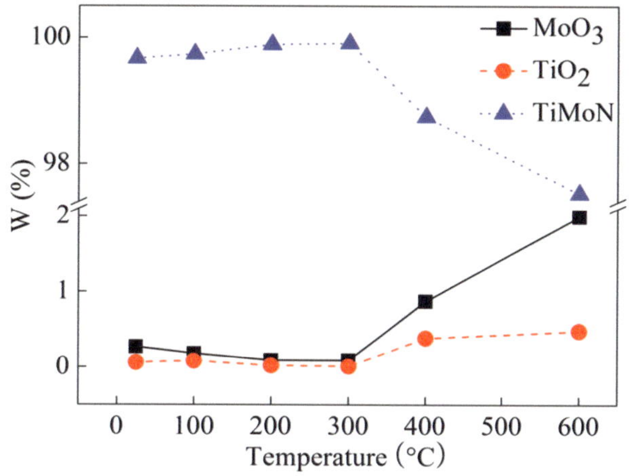

Figure 6.
The relative mass fraction of Ti-Mo-N, MoO_3, and TiO_2 on the wear tracks of Ti-Mo-N films at testing temperatures.

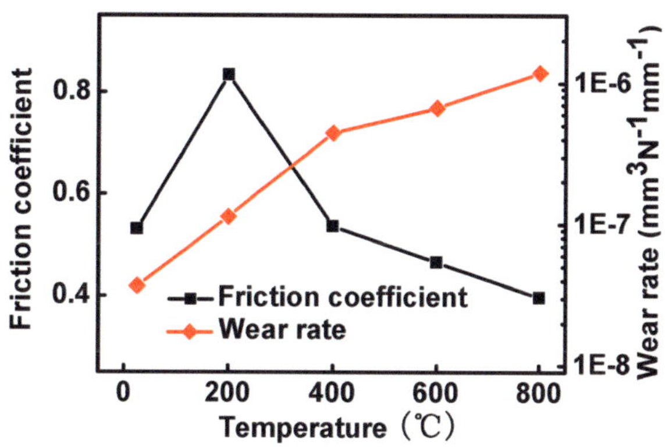

Figure 7.
The friction coefficient and wear rate of Ti-W-N composite films at different temperatures.

400°C, the Magnéli phase MoO_3 and oxidized phase TiO_2 appeared contributes to the decrease of average friction coefficient, and the relative mass fraction of the friction phase in the film wear marks increases gradually with the increase of ambient temperature, which can lubricate the wear marks and friction pairs and make the interaction between them tend to ease, so the average friction coefficient of the film decreases gradually with the increase of ambient temperature. However, the MoO_3 layers are only combined by van der Waals force. MoO_3 is very diffusive and easy to be worn by counterpart during the wear test, so MoO_3 is not wear-resistant although it can decrease the friction coefficient.

In addition, there are similar conclusions in Ti-W-N composite film. TiN incorporating tungsten can form the substitutional solid solution of (Ti, W) N as well and precipitate W_2N when the content of tungsten reaches a specific value. **Figure 7** illustrates the friction coefficient and wear rate of Ti-W-N composite films at 35.29 at.% W under different temperatures. When the temperature rises to 400°C, the friction coefficient decreases sharply because the TiWN composite film is oxidized, and the oxidation phases of Ti and W are formed on the surface of the film, which means TiO_2 and WO_3. WO_3 with layered structure plays a role in the lubricant under the high-temperature friction. When the content of WO_3 increases, the coefficient of friction decreases. WO_3

has a melting point of about 730°C, and when the temperature continues to rise to 800°C, the melted WO_3 has better lubricating properties than the dry friction, which leads to the further decrease of friction coefficient. Beside this, the friction coefficient will keep a low value due to the constant formation of liquid WO_3 during the friction process.

5. Conclusion

1. The Magnéli phase exhibits the excellent lubricating properties at high temperatures; however, it could be worn away easily by the counterpart during the wear test due to its weakly bonded lattice planes.

2. Addition of aluminum and silicon into the TMN-based films containing Magnéli elements could improve the thermal stability and decrease in the content of Magnéli phase on the wear track surface. It could enhance the wear resistance property to some extent.

Author details

Hongbo Ju
School of Materials Science and Engineering, Jiangsu University of Science and Technology, Zhenjiang, Jiangsu Province, China

*Address all correspondence to: hbju@just.edu.cn

References

[1] Ozsdolay BD, Mulligan CP, Balasubramanian K, Huang L, Khare SV, Gall D. Cubic β-WNx layers: Growth and properties vs N-to-W ratio. Surface and Coating Technology. 2016;**304**:98-107

[2] Ozsdolay BD, Mulligan CP, Guerette M, Huang L, Gall D. Epitaxial growth and properties of cubic WN on MgO(001), MgO(111), and Al2O3(0001). Thin Solid Films. 2015;**590**:276-283

[3] Reeswinkel T, Sangiovanni DG, Chirita V, Hultman L, Schneider JM. Structure and mechanical properties of TiAlN–WNx thin films. Surface and Coating Technology. 2011;**205**(20):4821-4827

[4] Mu Y, Liu M, Zhao Y. Carbon doping to improve the high temperature tribological properties of VN coating. Tribology International. 2016;**97**:327-336

[5] Erdemir A. Review of engineered tribological interfaces for improved boundary lubrication. Tribology International. 2005;**38**(3):249-256

[6] Cai Z, Pu J, Lu X, Jiang X, Wang L, Xue Q. Improved tribological property of VN film with the design of pre-oxidized layer. Ceramics International. 2019;**45**(5):6051-6057

[7] Ju H, N Ding JX, et al. Crystal structure and the improvement of the mechanical and tribological properties of tungsten nitride films by addition of titanium. Surface and Coating Technology. 2018;**345**:132-139

[8] Guo H, Han M, Chen W, Lu C, Li B, Wang W, et al. Microstructure and properties of VN/Ag composite films with various silver content. Vacuum. 2017;**137**:97-103

[9] Ju H, Jia P, Xu J, Yu L, Asempah I, Geng Y. Crystal structure and high temperature tribological behavior of niobium aluminum nitride films. Materialia. 2018;**3**:202-211

[10] Ju H, P Jia JX, et al. The effects of adding aluminum on crystal structure, mechanical, oxidation resistance, friction and wear properties of nanocomposite vanadium nitride hard films by reactive magnetron sputtering. Materials Chemistry and Physics. 2018;**215**:368-337

[11] Hazra B, Baranwal P, Bera S, Show BK. Improvement in dry sliding wear resistance of Al-17Si-5Cu alloy after an enhanced heat treatment process. Transactions of Nonferrous Metals Society of China. 2018;**28**(9):1705-1713

[12] Yang JF, Yuan ZG, Liu Q, Wang XP, Fang QF. Characterization of MoAlN nanocrystalline films synthesized by reactive magnetron sputtering. Materials Research Bulletin. 2009;**44**(1):86-90

[13] Ju H. Effects of Al and Si on Microstructure, Mechanical and Friction and Wear Properties of TiN and Mo_2N-based Films. Jiangsu University of Science and Technology; 2013

[14] Glatz SA, Koller CM, Bolvardi H, Kolozsvári S, Riedl H, Mayrhofer PH. Influence of Mo on the structure and the tribomechanical properties of arc evaporated Ti-Al-N. Surface and Coating Technology. 2017;**311**:330-336

[15] Yang Q, Zhao LR, Patnaik PC, Zeng XT. Wear resistant TiMoN coatings deposited by magnetron sputtering. Wear. 2006;**261**(2):119-125

[16] Londoño-Menjura RF, Ospina R, Escobar D, Quintero JH, Olaya JJ, Mello A, et al. Influence of deposition temperature on WTiN coatings

tribological performance. Applied Surface Science. 2018;**427**:1096-1104

[17] Cicek H, Baran O, Keles A, Totik Y, Efeoglu I. A comparative study of fatigue properties of TiVN and TiNbN thin films deposited on different substrates. Surface and Coating Technology. 2017;**332**:296-303

Chapter 11

Testing of the Resistance to Scuffing of Spiral Bevel Gears: Test Rig, Method, and Results of Verification Testing

Waldemar Tuszynski, Marek Kalbarczyk, Bartosz Kiser, Michal Michalak, Remigiusz Michalczewski, Jerzy Mydlarz, Witold Piekoszewski, Marian Szczerek and Jan Wulczynski

Abstract

In spite of long-term development of the technology of bevel gear production, the automotive industry reports various operational demands such as the need of the size and mass reduction of gears without lowering their durability and reduction of friction leading to a decrease in the energy losses and a decrease of the tendency to scuffing. What is more expected, EU regulations may impose the use of new generation gear oils (ecological-friendly) providing the proper operational properties of the transmission. In view of these demands, a new, bevel gear test rig and scuffing test method have been developed at ITeE-PIB. The idea resulted from a necessity to improve reliability of tests—popular gear tests are run mostly on spur gears having the tooth geometry significantly different than bevel gears. The test rig, test method, and results of verification testing are presented. The effect of various gear oils and the deposition of a low-friction coating on the resistance to scuffing were investigated. It is shown that the new test rig fulfils the research requirements and that the new test method has a good resolution.

Keywords: bevel gear, test rig, scuffing, gear oil, coating

1. Introduction

1.1 Types of automotive drivetrains

In modern motor vehicles, like cars, lorries, and buses, the design of the drivetrain depends on the function of a given vehicle. For small- and medium-sized cars, the automotive engineers pay attention to design such a drivetrain that occupies as small space as possible. In this case, the motor and gearbox are located crosswise in the front part of the vehicle; thus, it is the front wheels that are driven, and the functionality of the transmission, differential, and associated components of the driven axle is integrated into one assembly called the transaxle. This design is very compact; however, it limits the maximum motor power to approximately 150 kW, which is a result of offloading the front wheels when the vehicle accelerates.

A diagram of the described drivetrain is shown in **Figure 1** [1]. In this and the next few figures, the rectangle with the four circles represents the motor and its cylinders.

In the vehicles of a higher class, being larger in size, the criterion of the compactness of the drivetrain is not so important. In such vehicles, as well as in vans and lorries, the so-called classic drivetrain is employed. In this case, the motor is located longitudinally in the front part of the vehicle, and the rear wheels are driven (**Figure 2**) [1].

In the solution portrayed in **Figure 2**, the driving torque is transmitted from the motor via the clutch, gear box, and drive shaft to the axle with a final drive, differential, shafts, and finally to the wheels.

In motor vehicles having the classic drivetrain, during acceleration, the back wheels are loaded; thus, the problem of the adherence of the driven wheels to the

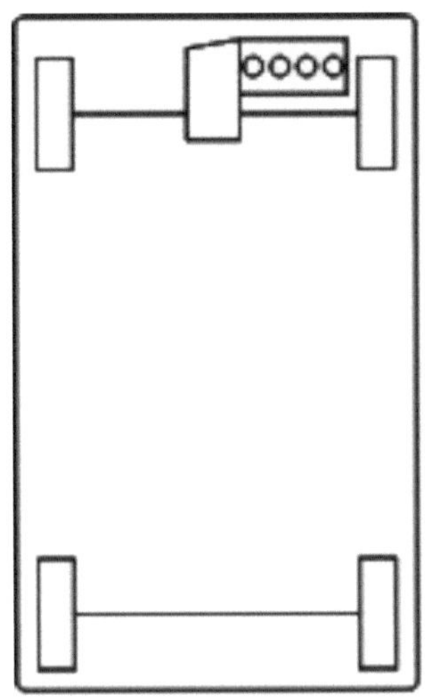

Figure 1.
Diagram of a transverse drivetrain [1].

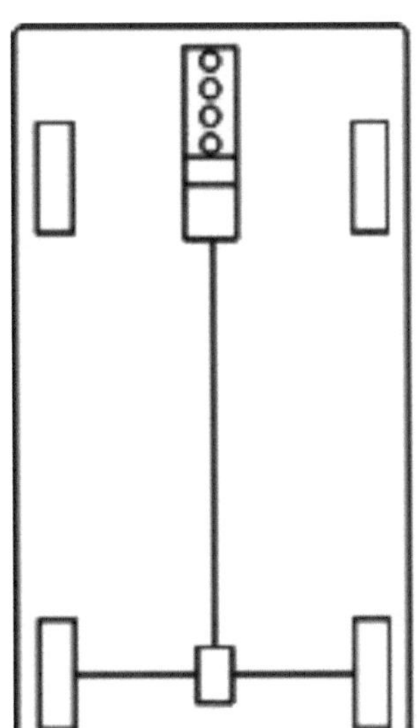

Figure 2.
Diagram of a classic drivetrain [1].

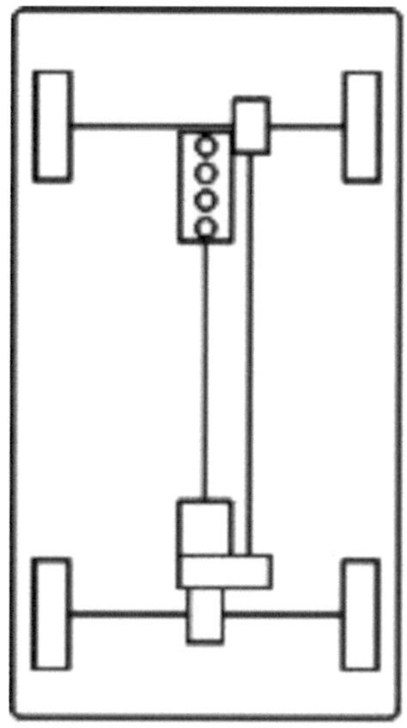

Figure 3.
Diagram of a drivetrain with all wheels driven [1].

road does not limit the motor power. It is very important in high-class cars (e.g. BMW, Mercedes), as well as in lorries.

The highest acceleration and off-road functionality are exhibited by vehicles having all wheels driven. In such vehicles, the number of final drives equals the number of the driven axles. A scheme of the "state-of-the-art" drivetrain used in, e.g. Nissan GT-R sports vehicles, is presented in **Figure 3** [1].

Unique features of this design are two driving shafts. The first one transmits the driving torque from the motor to the gearbox. Then, the torque is transmitted to the rear axle final drive (transaxle). The second driving shaft transmits the torque from the gearbox via the distribution box to the final drive of the front axle. This sophisticated design makes it possible to uniformly load both axles and achieve a neutral vehicle behaviour when turning.

Another example of a drivetrain that drives all the vehicle wheels is a famous Willys-Overland used by the US Army during the World War II. Its numerous "clones" are still produced in the world. It is interesting that its original version Willys-Overland is still employed by the Indian Army. The described drivetrain is shown in **Figure 4** [1].

The drivetrain shown in **Figure 4** is equipped with two axles. The driving torque produced by the motor is transmitted from the gearbox via the distribution box to the two driving shafts and then to the front and rear axles.

1.2 Final drives

A typical axle consists of the final drive, differential, and half shafts, which transmit the driving torque to the wheels.

The final drive is to transmit the driving torque to the wheels, at the same time, reducing the transmitted rotational speed, very often changing its direction, and increasing the driving torque. In most cases, the final drive consists of a pair of toothed gears.

The final drive can be one-stage, two-stage, and multistage, depending on the number of the reduction gears. Concerning the simplest, one-stage final drives for the automotive purposes, four of these types are applied (**Figure 5**) [2].

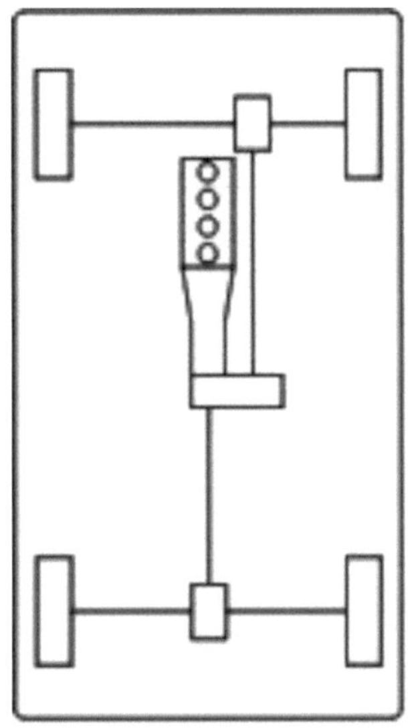

Figure 4.
Diagram of a drivetrain with all wheels driven used in the Willys-Overland off-road vehicles [1].

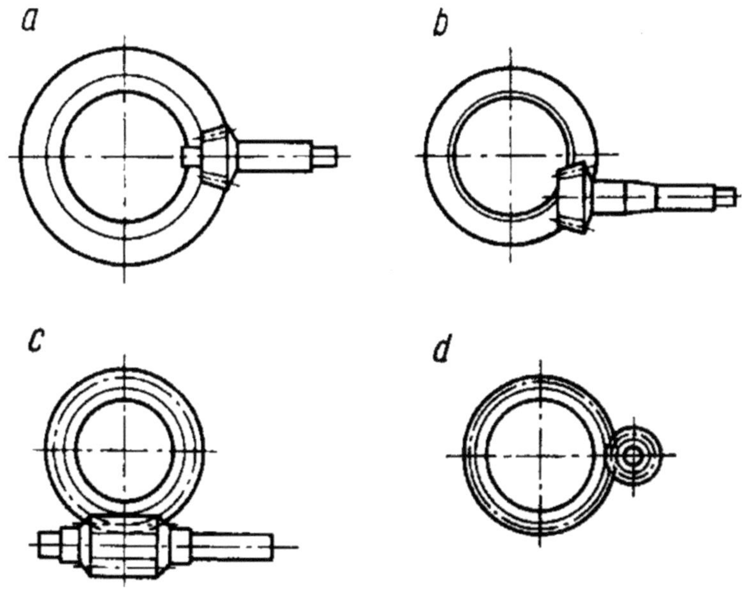

Figure 5.
The typical, one-stage final drives produced by the automotive industry: (a) spiral bevel gears (offset = 0), (b) hypoid gears (offset >0), (c) worm gears, and (d) helical gears [2].

A hypoid gear is a type of a spiral bevel gear whose axis does not intersect with the axis of the meshing gear. This is why the distance between these axes, called the offset, is higher than 0. In spiral bevel gears, the offset is exactly 0.

Spiral bevel gears or hypoid gears are used in the final drives of the drivetrains shown in **Figures 2–4**, while in the drivetrain presented in **Figure 1**, helical gears constitute the final drive.

Although nowadays, in vehicles, practically only helical gears (e.g. in the drivetrain from **Figure 1**) or hypoid gears (e.g. in the drivetrain from **Figure 2**) are used, spiral bevel gears are still very important, because they are broadly applied in lorries, special vehicles, and buses, where the axle size and the noise emission are

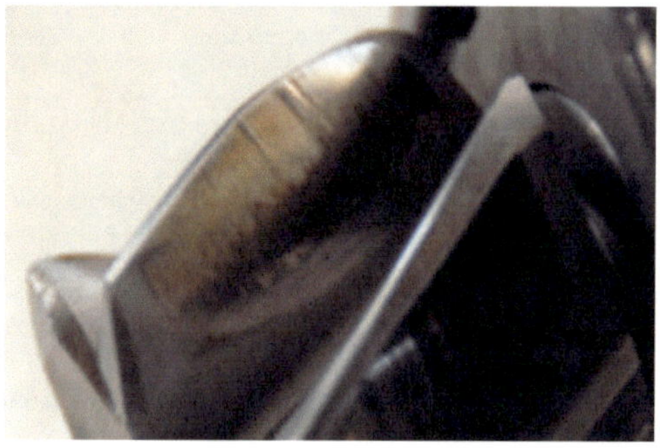

Figure 6.
Scuffed area on the tooth flank of a spiral bevel gear.

of minor importance but the gear efficiency is a priority. For example, spiral bevel gears are employed for the final drives in the drivetrain of the Willys-Overland off-road vehicles (**Figure 4**); their "clones" are still produced throughout the world.

It should be emphasised here that spiral bevel gears are used not only in motor vehicles. It can be estimated that, in case of, e.g. speed reducers with bevel-shaped gears, spiral bevel gears are applied in 90% of such devices, leaving only 10% to hypoid gears.

In spite of long-term development of the manufacturing technology and service of spiral bevel gears, there are still problems related to friction. The following problems, coming from excessive friction in such gears, still exist: an increase in the oil temperature followed by a rise in the tendency to scuffing. A typical scuffed area on the flank of the spiral bevel gear tooth is shown in **Figure 6**.

For the purpose of research on spiral bevel gears, the authors have developed a new test rig and a method for testing scuffing of such gears. The new testing machine and the test method will be a subject of the body of this chapter.

1.3 Automotive gear lubrication

Comprehensive characteristics of the automotive gear oils were provided in the former authors' chapter, being a part of the book [3]. Generally, the characteristics of the automotive gear oils were based on the source document provided by the American Petroleum Institute (API) [4].

However, to help the reader to analyse the data in this chapter, the authors decided to repeat information given in the book [3].

As in the book [3], emphasis was put on the classification of manual transmission fluids (MTF), i.e. automotive gear oils, excluding oils for automatic transmissions, called the automatic transmission fluids (ATF) and excluding oils for off-road vehicles (e.g. tractors). For simplification, manual transmission fluids (MTF) will be called the equivalent name—"automotive gear oils"—that will be used throughout the text.

There are two different classifications of automotive gear oils.

The first one specifies lubricant service designations or the "performance levels" of automotive gear oils. These have been provided by the American Petroleum Institute (API) [4]. The API classification divides automotive gear oils into seven performance levels. Four performance levels are in current use, three are not. The reason for the performance level not to be in current use results from changes of manufacturers' recommended tests or the unavailability of test equipment, and it does not mean that such products have been withdrawn from the market.

The API classification is described in **Tables 1** and **2**.

API service designation	Application areas and short characterisation
GL-1	Manual transmissions operating under such mild conditions that straight petroleum or refined petroleum oil may be used satisfactorily Not satisfactory for many passenger car manual transmissions GL-1 oils may contain oxidation and rust inhibitors, defoamers, and pour depressants. Friction modifiers (FM) and extreme pressure (EP) additives shall not be used in GL-1 oils
GL-4	Axles with spiral bevel gears operating under moderate to severe conditions of speed and load or axles with hypoid gears operating under moderate speeds and loads GL-4 oils may be used in selected manual transmission and transaxle applications where MT-1 lubricants are unsuitable GL-4 oils contain up to 4% of extreme pressure (EP) additives
GL-5	Gears, particularly hypoid gears, in axles operating under various combinations of high-speed/shock-load and low-speed/high-torque conditions GL-5 oils contain up to 6.5% of extreme pressure (EP) additives
MT-1	Nonsynchronised manual transmissions used in buses and heavy-duty lorries API MT-1 does not address the performance requirements of synchronised transmissions and transaxles in vehicles and heavy-duty applications API MT-1 oils provide protection against the combination of thermal degradation, component wear, and oil seal deterioration, which is not provided by lubricants in current use meeting only the requirements of API GL-1, 4, or 5

Table 1.
API service designations (performance levels) of automotive gear oils in current use, according to the source document—API Publication 1560 [4].

API service designation	Application areas and short characterisation
GL-2	Automotive worm-gear axles operating under such conditions of load, temperature, and sliding velocities that lubricants, satisfactory for API GL-1 service, will not suffice GL-2 oils contain antiwear or film strength improvers specifically designed to protect worm gears Currently, it is very difficult to find products designed as "GL-2" in the automotive market
GL-3	Manual transmissions operating under moderate to severe conditions and spiral bevel axles operating under mild to moderate conditions of speed and load GL-3 oils provide load-carrying capacities exceeding those satisfying API GL-1 but below the requirements of API GL-4 oils GL-3 oils are not intended for axles with hypoid gears They contain up to about 3% of antiwear (AW) additives
GL-6	Hypoid gears designed with a very high pinion offset Their EP properties are typically better than of GL-5 oils GL-6 oils contain up to 10% of extreme pressure (EP) additives

Table 2.
API service designations (performance levels) of automotive gear oils not in current use, according to the source document—API Publication 1560 [4].

Apart from the designations from **Tables 1** and **2**, there is also one denoted as API GL-5(LS) or GL-5+. Gear oils that meet the requirements of this class contain special additives, called friction modifiers (FM), preventing the stick–slip phenomenon under conditions of limited slip (LS). As such, GL-5(LS) oils are intended for lubrication of limited slip differentials.

To reduce the number of various gear oils in the market and, in turn, simplify oil selection, many lubricant manufacturers implement more universal ("multi-class") gear oils. In this group, gear oils denoted as API GL-4/GL-5 or GL-4+ predominate.

SAE viscosity grade	Maximum temperature for viscosity of 150,000 cP [°C]	Kinematic viscosity at 100°C, cSt [mm^2/s] Minimum	Kinematic viscosity at 100°C, cSt [mm^2/s] Maximum
70 W	−55	4.1	—
75 W	−40	4.1	—
80 W	−26	7.0	—
85 W	−12	11.0	—
80	—	7.0	<11.0
85	—	11.0	<13.5
90	—	13.5	<18.5
110	—	18.5	<24.0
140	—	24.0	<32.5
190	—	32.5	<41.0
250	—	41.0	—

Table 3.
Automotive gear oil viscosity classification according to the source document—SAE J306:2005 [5].

The second classification was provided by the Society of Automotive Engineers (SAE) [5]. It is a viscosity classification. It divides automotive gear oils into 11 grades based on their rheological properties (**Table 3**).

For lubrication of automotive gears, multiviscosity grade oils are mostly employed, e.g. SAE 80W-90. This designation means that such an oil meets the requirements of both low-temperature (SAE 80W) and high-temperature grades (SAE 90).

1.4 Scuffing

The comprehensive analysis of the scuffing phenomenon was provided in the former authors' chapter, being a part of the book [6]. However, to help the reader analyse this chapter, the authors decided to extract the most important information and present it here.

1.4.1 Some definitions of scuffing

Despite long tradition of research on scuffing, its terminology has not been systemised so far. The term scuffing has many synonyms, e.g. seizure, scoring, galling, and seizing.

The variety of terms presented here makes interpretation of literature data difficult. Additionally, many terms are very often used as equivalents, e.g. scuffing and seizure.

The lack of consensus on the origin of scuffing as well as its symptoms is reflected by numerous definitions of the phenomenon. Some of them associate scuffing with wear, the others—with friction. The former are the definitions given or quoted by, e.g. Enthoven and Spikes, Dyson, Ludema, and Sadowski. Among the latter, one can classify the definition given by Nosal.

Enthoven and Spikes [7] have quoted the definition suggested by the Organisation for Economic Co-operation and Development, characterising scuffing as "localized damage caused by the occurrence of solid-phase welding between sliding surfaces, without local surface melting".

Dyson [8] has adopted the definition presented by The Institution of Mechanical Engineers: scuffing is "gross damage characterized by the formation of local welds between the sliding surfaces".

Ludema [9] has considered scuffing to be "a roughening of surfaces by plastic flow whether or not there is material loss or transfer".

Sadowski [10] has interpreted scuffing as "an interference of stabilised wear and a boundary example of such wear".

However, according to Nosal [11], scuffing is "a collection of phenomena taking place in the sliding pair, localized mainly deep inside the surface layer, producing an increased and unstabilised friction which is likely to result in seizure".

1.4.2 What does initiate scuffing?

Practical lubrication in most of the sliding pairs, e.g. high-speed toothed gears, is neither a purely hydrodynamic (HD) nor an elastohydrodynamic (EHD) process. Despite either HD or EHD, film supports most of the loading applied to the tribosystem elements, collisions between the highest surface asperities cannot be excluded. Therefore, besides fully lubricated friction, dry and boundary friction may also appear. Their common action is called the mixed friction or mixed lubrication [12].

The lubricating film, being produced at mixed friction, compared to such created at fully lubricated friction, is significantly thinner. Collisions of surface asperities give rise to a local load increase, then the lubricating film collapses, and scuffing takes place. It is a serious practical problem, because scuffing can appear in tribosystems creating apparently good conditions for lubrication.

A kinetic model, proposed by Nosal [11], presents scuffing to be a sequence of the following phenomena: collapse of the lubricating film ⇨ removal of oxides layers in microareas of contact, leading to direct metal–metal contacts ⇨ appearance of adhesive bonds due to a temperature increase or plastic deformation ⇨ development of adhesive bonds deeper and deeper inside the surface layer ⇨ shearing of adhesive bonds, tearing out and transfer of metal particles from one element onto the other, due to their relative movement ⇨ rapid development of the above phenomena ⇨ macroscopic range of destruction (scuffing).

The kinetic model of scuffing assumes that the transition from stabilised friction to scuffing requires some time, rejecting the idea that it takes place very rapidly, i.e. during $t \rightarrow 0$.

Scuffed areas on the tooth flank of a spiral bevel gear were presented earlier in **Figure 6**.

It is very important to differentiate between the terms scuffing and scoring. Scuffing has a physical nature and is a result of shearing adhesive bonds, which appear between the rubbing surfaces. Scoring has a rather mechanical nature and appears at very high loads. Under such conditions abrasion of the surface takes place—by the action of the very hard wear particles [13].

1.5 Some reasons for the bevel gear testing

In spite of long-term development of the technology of bevel gear production, the automotive industry reports various operational demands. The main one concerns the need of the size and mass reduction of gears without lowering their durability. The second demand relates to the reduction of friction, leading to the decrease in heat emission, lowering the energy losses, and above all, resulting in a decrease of the tendency to scuffing. The abovementioned effect can be achieved by the deposition of a thin, hard, and low-friction coating on the operating surface of the gears, although the present-day research is limited to spur gears [14–22].

The third important aspect is environmental protection, which is addressed by the use of new generation gear oils (ecological-friendly) providing the proper operational properties of the gear. In addition, the new EU law requirements, concerning the biodegradability of lubricants in various fields, are expected to be implemented soon. This is why numerous research investigations are now being conducted, which consider the application of ecological oils for various technological areas [23–26].

Apart from the demands mentioned above, an equally important issue is the diversification of automotive gear oils in respect to API GL performance levels. Present research concerns the simple specimens [27–29], such as the four-ball tribosystem, as well as spur gear investigation [30, 31]. However, there is a striking lack of research performed in this direction on spiral bevel gears.

The mentioned demands could be fulfilled only by a new tribotester, intended for testing bevel gears. There are some tribotesters of this type. The first one is the hypoid/bevel gear test rig [32, 33] developed in the Gear Research Centre (FZG) at the Technical University of Munich. The second one is the test rig [34] designed at NASA Glenn Research Center. This machine is intended for aeronautics purposes. However, these test rigs are not widely available.

Poor availability of bevel gear test rigs is probably the reason why most of the publications in this area focus on the two aspects:

- Optimisation of the geometry of bevel gears, optimisation of the conditions of meshing teeth of bevel gears, as well as improvement of the production technology—by using the tools of mathematical modelling [35–38]

- Case studies, focusing on the explanation of operational causes of bevel gear scuffing, surface fatigue wear (pitting), and teeth breakage in various vehicles, machines, and devices, i.e. lorry transmission, turbojet gearboxes, or in the transmissions of industrial conveyors [39–41]

In spite of the test rig, for the investigation of bevel gears, a necessary factor is the usage of proper test methods. Unfortunately, a review of the most common standards indicates that there is a complete lack of any standardised methods for testing scuffing of bevel gears. The existing standards concern the means of construction calculations for providing the proper resistance to scuffing (ISO/TR 13989-1,2), and it has to be underlined that these calculation methods are treated as controversial. For this reason, the American Gear Manufacturers Association (AGMA) has not published such a method as a standard yet. The lack of research in this area is clearly visible.

In contradiction to the calculation methods, available as standards or computer programmes, the proposals of authorial methods for friction and wear testing of hypoid or bevel gears are presented very rarely in publications [32, 33, 42, 43].

As it was mentioned before, answering to the issues specified above at the Tribology Department of the Institute for Sustainable Technologies (ITeE-PIB), a new test rig for the investigation of bevel gears was developed, and the authorial method for testing scuffing of bevel gears was designed. The testing machine and the test method are the subjects of the chapter.

2. Bevel gear test methodology

2.1 New bevel gear test rig

The kinematic scheme of the developed test rig is presented in **Figure 7**. In the line of testing equipment for tribological research, developed at ITeE-PIB, the bevel gear testing machine is denoted as T-30.

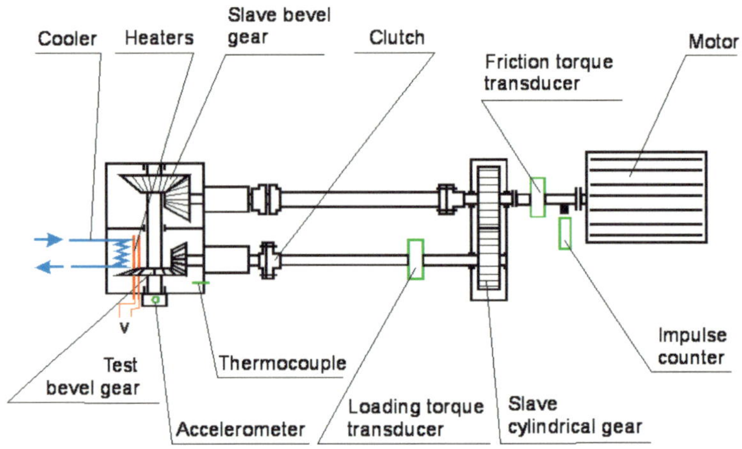

Figure 7.
Kinematic scheme of the T-30 back-to-back bevel gear test rig.

The main part of the T-30 test rig consists of three gears—a test bevel gear, slave bevel gear, and slave cylindrical (helical) gear—and in the figure, all the teeth were presented as if they were straight for simplification. The test bevel gear consists of a pinion and a gear wheel, which are immersed in the investigated oil. Before the test, the test chamber is preheated to the required temperature by means of a heating unit. The test chamber is equipped with cooling channels (in **Figure 7** called the cooler) connected with a heat exchanger working in a closed water circulation system. This construction makes it possible to cool the oil before the gear inspection. Furthermore, it extends the application area of the T-30 test rig for investigation under conditions of constant oil temperature (necessary for surface fatigue testing).

The testing machine is equipped with a circulating power system. After the application of the required load and fixing the shafts and the clutch, the torque circulates in the shaft-gear system. The load (shaft torsion) is applied by means of the lever unit with the number of weights, providing the possibility of obtaining a torque up to 720 Nm. The applied torque and its changes during the test are measured by means of the loading torque transducer.

The measurement of friction torque is realised with the use of a friction torque transducer mounted between the motor and the slave cylindrical gears. The purpose of using friction torque measurement is to make it possible to compare materials from which the test gears are produced from the point of view of a possible friction reduction. The motor is supplied through a power inverter, providing the possibility of speed regulation.

A photograph of the T-30 bevel gear testing rig, equipped with the control and measuring system, is presented in **Figure 8**.

The control system makes it possible to regulate the motor speed to change the lubrication conditions of the tested gear by changing the motion direction, to regulate the temperature of the oil, and to automatically turn off the motor after reaching the required number of test pinion revolutions or after exceeding the selected level of vibrations. The main elements of the control system are mounted in the measuring cupboard (**Figure 8**—left side).

The measuring system (**Figure 8**) consists of a digital amplifier, the set of measuring transducers, and a computer equipped with specialised software. During the test, the following quantities are measured and registered: the loading torque between the gears, the friction torque, the motor speed, the number of test pinion revolutions, the temperature of the tested oil, the level of vibrations, and the test duration.

Figure 8.
Photograph of the T-30 back-to-back bevel gear test rig with its control and measuring system (on the left) and water circulating cooling system (on the right) [44].

The measured signals from the transducers are amplified and transferred to the computer. The specialised software developed at ITeE-PIB makes it possible to display the curves of measured values in relation to test duration. At the end of the test, the measurements are saved on the hard disc, and then research reports can be printed.

The new test rig was designed with the cooperation between ITeE-PIB and Invenio Ltd. and manufactured by ITeE-PIB.

2.2 New bevel gear scuffing test

A simplified scheme of the geometric configuration of the test bevel gears is presented in **Figure 9**.

The pinion and the wheel have a spiral line of teeth. The direction of motion shown in **Figure 9** is defined by the authors as "normal" viewed from the test chamber's front cover, and the wheel rotates counterclockwise. From the same point

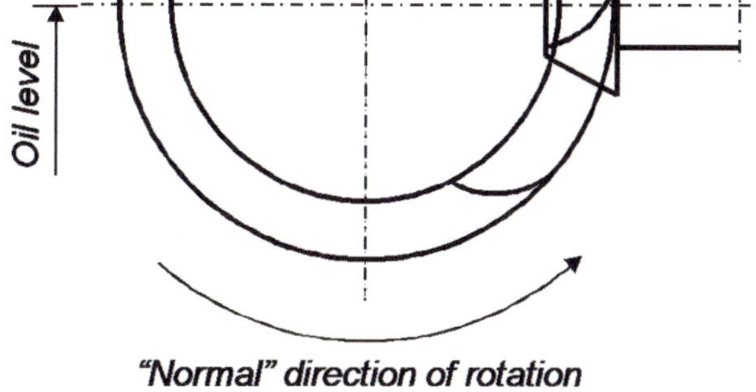

Figure 9.
Simplified scheme of the geometric configuration of the test bevel gears [44].

of view, the pinion is mounted behind the wheel, which was skipped in **Figure 9** in order to simplify the scheme.

Under the load action direction, which results from the test rig construction, the convex flank of the pinion tooth is pressed to the concave flank of the wheel. As a consequence, the convex flank of the pinion is recognised as the investigated one. The pinion and the wheel are immersed in the tested oil up to the level of the gear axis.

The developed test method consists in performing research on the lubricated spiral bevel gears, operating under the conditions presented below, at the constant rotational speed, gradually increasing load, with a starting temperature the same for every test, until the failure load stage (FLS) is reached, which is a basic indicator of the resistance to scuffing, or until the 12th load stage is reached. FLS is the load stage under which the area, which is damaged by polishing, scratches, and scoring, exceeds the area of one pinion tooth (≥ 450 mm^2). Therefore, the FLS is a measure of the anti-scuffing properties of the investigated oil or the scuffing resistance of the coating deposited on one or both test gear elements.

Test conditions

• Type of test gears	Bevel gears with spiral teeth line
• Rotational speed of the motor	3000 rpm
• Average circumferential speed	7.7 m/s
• Single run duration	15 min
• The motion rotational direction	"Normal" (**Figure 9**)
• Min. and max. Load stage	From 1 to 12 (the 13th and 14th stage is allowed)
• Loading torque	From 3 to 535 Nm, gradually increasing
• Hertzian contact stress	From 0.1 to 1.5 GPa
• Oil starting temperature	90°C (not stabilised during the test)
• Type of lubrication	Immersive (approx. Volume of oil: 2 l)
• Level of oil	To the axis of the test gear elements

Depending on the requirements, beside the basic scuffing resistance indicator, which is FLS, for evaluation of anti-scuffing properties, an additional criterion can be used, which is the damage progression of the pinion teeth flanks. For this purpose, during every pinion inspection, apart from the total damaged area, the most common mode of wear is also notified in the report sheet—according to **Table 4**.

When testing a low-friction hard coating deposited on one of the test gears, the coated element "polishes" the uncoated one due to a mild abrasive action of the coating surface. Therefore, polishing areas on the teeth flank of the uncoated gear can be neglected when inspecting test gears for the sings of damage.

2.3 Test spiral bevel gears

The pair of test spiral bevel gears is shown in **Figure 10**.

Mode of wear	Symbol	Appearance
Polishing	W	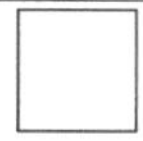
Scratches	R	
Scoring	B	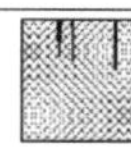
Scuffing	Z	

Table 4.
Possible modes of wear of the test pinion.

Figure 10.
Test spiral bevel gears.

The spiral bevel gears were manufactured in a high, fifth grade of precision, defined in DIN 3965. They were made of 18CrNi8 steel. The material was carburised (0.6–0.9 mm in depth). Then it was tempered to achieve the hardness of 56–60 HRC. Machining was performed using Klingelnberg's method.

Technical specifications of the test spiral bevel gears are presented in **Table 5**.

2.4 Oils tested

For the research, the following oils were used: the commercial mineral-based oils of API GL-1 and GL-5 performance level, intended for vehicle mechanical gears, and VG 220 viscosity grade, ecological oil, intended for industrial gears (this oil has been developed at ITeE-PIB). The characteristics of the investigated oils are presented in **Table 6**.

2.5 Materials tested

Two material configurations were tested. The first one consisted of a steel pinion and steel wheel, and the second one consisted of a steel pinion and a-C:H:W coated steel wheel. The photographs of both configurations are presented in **Figure 11**.

The a-C:H:W coating is a multilayer coating. The structure of the coating is presented in **Figure 12**, showing the results of the depth profile quantitative analysis obtained by a glow discharge optical emission spectrometer (GDOES). The reference sample was deposited on the Armco Fe substrate material (technically pure Fe).

As it is shown in **Figure 12**, on the Fe-based substrate, the layer of Cr is deposited to increase adhesion of the tungsten carbide (WC) layer to the substrate. The hard WC is of crucial importance to achieve the high load-carrying capacity, and in case when the outer layer wears out, the WC layer protects the substrate material from abrasive wear.

Description	Pinion	Wheel
Manufacturing process	Klingelnberg	
Material	18CrNi8 steel	
Heat treatment	Case hardening	
Case depth	0.6–0.9 mm	
Surface hardness	56–60 HRC	
Core hardness	33–45 HRC	
Gear quality	5 according to DIN 3965	
Number of teeth	7	18
Normal module	8.70 mm	
Normal pressure angle	20°	
Mean spiral angle	35°	
Shaft angle	90°	
Working tooth flank	Convex	Concave

Table 5.
Main material and geometrical characteristics of the test spiral bevel gears.

Symbol	API GL service designation	Viscosity grade	Type of base oil	Application and short characteristics
GL-1 90	GL-1	SAE 90	Mineral	Manual transmissions operating under such mild conditions that straight petroleum or refined petroleum oil may be used satisfactorily GL-1 oils do not contain any lubricating additives
GL-5 80W-90	GL-5	SAE 80W-90	Mineral	Gears, particularly hypoid gears, in axles operating under various combinations of high-speed/shock-load and low-speed/high-torque conditions GL-5 oils contain up to 6.5% of extreme pressure (EP) additives
Eko VG 200	—	VG 220	Natural (vegetable)—mixture of rapeseed and castor oils	Industrial transmissions and transmissions in machine tools and in machines working especially in the paper industry The oil contains 2% (wt.) of sublimated sulphur, performing as an EP additive

Table 6.
Characteristics of the tested gear oils.

The outer layer is responsible for the interaction with the mating element of the tribosystem. It is a multilayer structure (colloquially named WC/C) that consists of WC-reach DLC layers with DLC interlayers. The presence of nickel is required for technological reasons. Low-friction properties are a result of a high concentration of amorphous carbon (sp^2) in the outer layer.

Figure 11.
Material configurations of the test bevel gears: (a) steel pinion, steel wheel, and (b) steel pinion, a-C:H:W coated wheel.

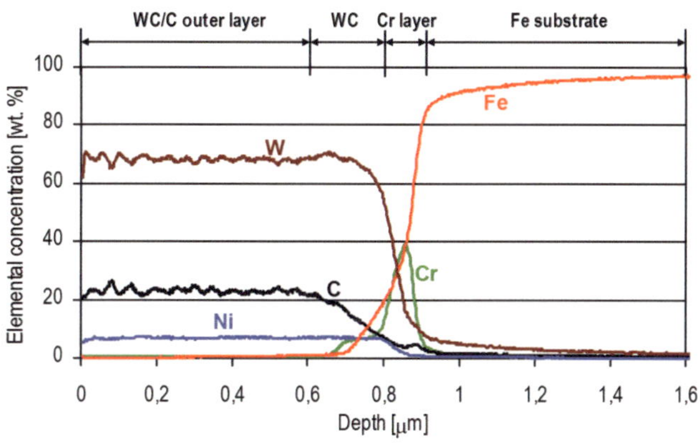

Figure 12.
Structure of the a-C:H:W coating; the results of the quantitative GDOES depth profiling (hydrogen was not analysed).

3. Results of verification testing of the T-30 bevel gear test rig

In **Figure 13**, the FLS values obtained for investigated oils and material configurations are presented. The expected repeatability intervals are added to the bars.

The failure load stage result analysis indicates that the resolution of the scuffing test method is very high—the FLS differences between all the tested cases are statistically significant. It is also of great importance that, for all the tested cases, the 12th load stage was not exceeded. The maximal obtained FLS level is not higher than 10, which is an additional confirmation of the very high resolution of the developed test method.

In the steel pinion/steel wheel configuration, the GL-1 90 oil without additives exhibits low resistance to scuffing (represented by very low—3rd—FLS), whereas the GL-5 80 W-90 oil, with significant concentration of EP additives, makes it possible to reach much higher FLS value (eighth). The Eko VG 220 oil is characterised by an intermediate value of failure load stage, which indicates that it can be used as a lubricant only for bevel gears, which operate under conditions of low and moderate loads, still providing a measurable ecological effect.

According to the facts presented by the researchers [17–22] who performed scuffing tests on spur gears, there is a possibility to improve the resistance to scuffing of gear elements by deposition of a thin, hard coating, i.e. a-C:H:W, on at least one of the gears. This thesis is fully confirmed by the results of the

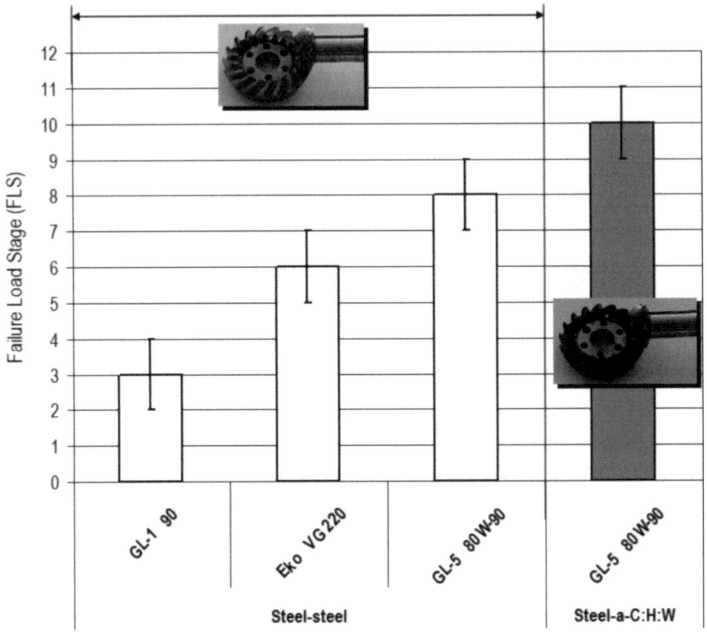

Figure 13.
Failure load stage (FLS) for the tested gear oils and material configurations.

bevel gear scuffing test performed on the configuration of a steel pinion mating a-C:H:W coated steel wheel for which the highest value of FLS was obtained (tenth). This test result indicates that the thin hard coatings represent a very high application potential for scuffing resistance improvement in the area of heavily loaded gears.

As stated before, considering the assessment of the anti-scuffing properties of oils and the resistance to scuffing of various material combinations of gear elements, in particular with coatings, one more important criterion can be used, which is the progression of wear, occurring on the flanks of pinion teeth under stepwise increasing load.

The detailed analyses comprised the modes of wear that occurred on most of the pinion teeth. The results of the analyses are compiled in **Table 7**. Below the symbols of the wear mode, the total damaged area of the pinion teeth is presented, expressed in mm^2. The description of wear modes is provided earlier in **Table 4**.

It can be observed that, as the load increases, the progression of tooth wear is the fastest for the gears lubricated with GL-1 90 oil, because the scuffing is initiated just after the test at the first load stage.

When considering the wear modes during two tests on two different material combinations lubricated with GL-5 80 W-90 oil, the eighth load stage causes the scuffing of the steel-steel pair, while on the steel pinion meshing with a-C:H:W coated wheel, only scoring (in a form of groves) appears. It is because the presence of the coating causes a lower level of affinity than in the steel-steel case. Additionally, the high hardness of the coating, decreasing the tendency to adhesive bonding, significantly mitigates scuffing. As an effect of these factors, the steel-coating material configuration reached a high tenth load stage.

Thus, the additional criterion (i.e. progression of wear on the teeth flanks) constitutes a very useful supplement to the basic criterion, which is the failure load stage, and it can provide additional and valuable information on the damage process of the teeth.

Concerning the coating-steel material combination of bevel gear teeth, it is very difficult to compare the wear results reported in this chapter with the results obtained by other researchers, because such papers are extremely rare. For example,

DOI: http://dx.doi.org/10.5772/intechopen.84848

Load stage	Steel—steel 			Steel—a-C:H:W 
	GL-1 90	**Eko VG 200**	**GL-5 80 W-90**	**GL-5 80 W-90**
1	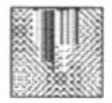 216 mm²	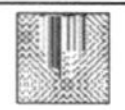 ≈0	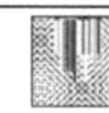 ≈0	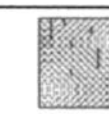 ≈0
2	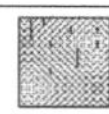 295 mm²	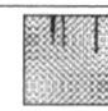 7 mm²	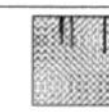 ≈0	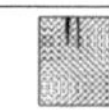 ≈0
3	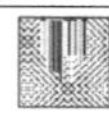 910 mm²	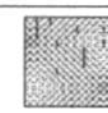 16 mm²	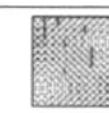 ≈0	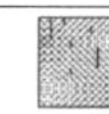 ≈0
4		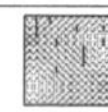 31 mm²	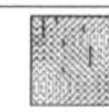 ≈0	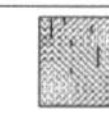 ≈0
5		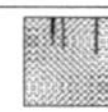 54 mm²	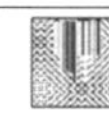 ≈0	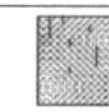 ≈0
6		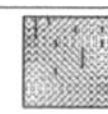 1225 mm²	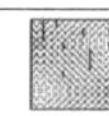 70 mm²	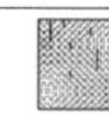 ≈0
7			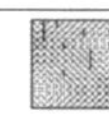 116 mm²	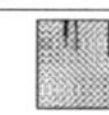 72 mm²

Load stage	**Steel—steel** 			**Steel—a-C:H:W** 
	GL-1 90	**Eko VG 200**	**GL-5 80 W-90**	**GL-5 80 W-90**
8			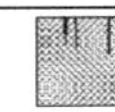 1575 mm^2	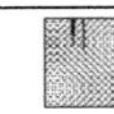 130 mm^2
9				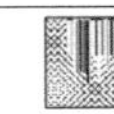 164 mm^2
10				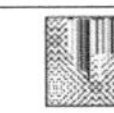 960 mm^2

Table 7.
The wear modes found on the teeth of the test pinion at increasing stages of load; the research was performed using various gear oils and material combinations.

Basiniuk et al. [45] studied the behaviour of coated bevel gears, however, from a point of view of a noise and friction reduction.

The authors plan to continue their research on the coated bevel gears. The aim will be to identify mechanisms of interaction between the surface and the gear oil of different chemistry in the steel-coating and coating-coating friction zones of the meshing teeth. Scientific publications, relating to this aspect, concern mainly testing on simple model specimens. The most recent papers in this field are, e.g. [46–58].

4. Conclusions

Based on the carried out research, the following conclusions can be drawn:

- The developed method for testing scuffing of spiral bevel gears exhibits a good resolution.

- It is possible to improve the resistance to scuffing by the application of a thin, low-friction coating on one of the elements of the spiral bevel gear.

The developed test rig with the test method can be used for the assessment of the anti-scuffing properties of oils intended for bevel gear lubrication and also for the determination of the resistance to scuffing of bevel gears from the perspective of materials used for gears and possibly the tooth surface processing. Thus, they can be applied in R&D industrial laboratories of lubricant producers, manufacturers of devices that contain bevel gears, automotive industry, as well in laboratories of technical universities where the aspects of bevel gear tribology and design are studied and investigated.

Acknowledgements

The research was financed by the National Centre for Research and Development (NCBiR) within the scope of Research Project No. N R03 0019 06.

Author details

Waldemar Tuszynski[1]*, Marek Kalbarczyk[1], Bartosz Kiser[2], Michal Michalak[1], Remigiusz Michalczewski[1], Jerzy Mydlarz[2], Witold Piekoszewski[1], Marian Szczerek[1] and Jan Wulczynski[1]

1 Tribology Department, Institute for Sustainable Technologies—National Research Institute (ITeE-PIB), Radom, Poland

2 Invenio Ltd, Tychy, Poland

*Address all correspondence to: waldemar.tuszynski@itee.radom.pl

References

[1] Mydlarz J. Model and analysis of diagonal transmission system for a car [thesis]. Poland: Silesian Institute of Technology Gliwice; 2011 (in Polish)

[2] Jaskiewicz Z. Drive Lines of Motor Vehicles—Driving Axles. Warsaw: WKL; 1977. (in Polish)

[3] Tuszynski W, Michalczewski R, Piekoszewski W, Szczerek M. Modern automotive gear oils—Classification, characteristics, market analysis, and some aspects of lubrication. In: Chiaberge M, editor. New Trends and Developments in Automotive Industry. Rijeka: InTech; 2011. pp. 297-322

[4] American Petroleum Institute (API). Lubricant Service Designations for Automotive Manual Transmissions, Manual Transaxles, and Axles. 7th ed. API Publication 1560; 1995

[5] Society of Automotive Engineers (SAE). Automotive gear lubricant viscosity classification. SAE. 2005:J306

[6] Burakowski T, Szczerek M, Tuszynski W. Scuffing and seizure—Characterization and investigation. In: Totten GE, Liang H, editors. Mechanical Tribology. Materials, characterization, and applications. New York: Marcel Dekker, Inc; 2004. pp. 185-234

[7] Enthoven J, Spikes HA. Infrared and visual study of the mechanisms of scuffing. Tribology Transactions. 1996;**39**:441-447

[8] Dyson A. Scuffing—A review. Tribology International. 1975;**8**:77-87

[9] Ludema KL. A review of scuffing and running in of lubricated surfaces with asperities and oxides in perspective. Wear. 1984;**100**:315-331

[10] Sadowski J. Thermodynamic Aspects of Tribological Processes. Radom: Radom Technical University; 1997. (in Polish)

[11] Nosal S. Tribological Aspects of Scuffing in Sliding Pairs. Poznan: Poznan Technical University; 1998. (in Polish)

[12] Hebda M, Wachal A. Tribology. Warsaw: WNT; 1980. (in Polish)

[13] Nadolny K. Gear Tribology. The Aspects of Durability and Reliability. Radom: ITeE; 1999. (in Polish)

[14] Kalin M, Vižintin J. The tribological performance of DLC-coated gears lubricated with biodegradable oil in various pinion/gear material combinations. Wear. 2005;**259**:1270-1280

[15] Martins RC, Moura PS, Seabra JO. MoS2/Ti low-friction coating for gears. Tribology International. 2006;**39**:1686-1697

[16] Martins R, Amaro R, Seabra J. Influence of low friction coatings on the scuffing load capacity and efficiency of gears. Tribology International. 2008;**41**:234-243

[17] Szczerek M, Michalczewski R, Piekoszewski W. The problems of application of PVD/CVD thin hard coatings for heavy-loaded machine components. In: Proceedings of ASME/STLE 2007 International Joint Tribology Conference; 22-24 October 2007; San Diego (USA)

[18] Michalczewski R, Piekoszewski W, Szczerek M, Tuszynski W. The lubricant-coating interaction in rolling and sliding contacts. Tribology International. 2009;**42**:554-560

[19] Tuszynski W, Michalczewski R, Szczerek M, Kalbarczyk M. A new scuffing shock test method for the determination of the resistance to

scuffing of coated gears. Archives of Civil and Mechanical Engineering (ACME). 2012;**12**:436-445

[20] Michalczewski R, Kalbarczyk M, Michalak M, Piekoszewski W, Szczerek M, Tuszynski W, et al. New scuffing test methods for the determination of the scuffing resistance of coated gears. In: Gegner J, editor. Tribology—Fundamentals and Advancements. Rijeka: Intech; 2013. pp. 187-215

[21] Michalczewski R, Piekoszewski W, Szczerek M, Tuszynski W, Antonov M. The rolling contact fatigue of PVD coated spur gears. Key Engineering Materials. 2013;**527**:77-82

[22] Michalczewski R, Kalbarczyk M, Piekoszewski W, Szczerek M, Tuszynski W. The rolling contact fatigue of WC/C-coated spur gears. Journal of Engineering Tribology. 2013;**227**(8):850-860

[23] Sulek MW, Wasilewski T. Antiseizure properties of aqueous solutions of compounds forming liquid crystalline structures. Tribology Letters. 2005;**18**:197-205

[24] Ilowska J, Gniady J, Kozupa M, Drabik J. Stabilization of biogreases made of rapeseed oils. Przemysl Chemiczny. 2011;**90**:1818-1822. (in Polish)

[25] Rogos E, Naraniecki B, Urbanski A, Lukosek M. Cutting fluids based on ethoxylates of the glycerol fraction from a biodiesel plant. Przemysl Chemiczny. 2011;**90**:1814-1815. (in Polish)

[26] Tuszynski W, Szczerek M, Michalczewski R, Osuch-Slomka E, Rogos E, Urbanski A. The potential of the application of biodegradable and non-toxic base oils for the formulation of gear oils - model and component scuffing tests. Lubrication Science. 2014;**26**:327-346

[27] Bisht RPS, Singhal S. A laboratory technique for the evaluation of automotive gear oils of API GL-4 level. Tribotest Journal. 1999;**6**(1):69-77

[28] Trzos M, Szczerek M, Tuszynski W. A study on the possibility of the Brugger test application for differentiation between the API-GL performance levels of gear oils. Archives of Civil and Mechanical Engineering (ACME). 2013;**13**:14-20

[29] Tuszynski W, Szczerek M. Qualitative discrimination between API GL performance levels of manual transmission fluids by comparing their EP properties determined in a new four-ball scuffing test. Tribology International. 2013;**65**:57-73

[30] Michaelis K, Höhn B-R, Graswald C. Scuffing tests for API GL-1 to GL-5 gear lubricants. In: Proceedings of 13th International Colloquium Tribology; 15-17 January 2002; Stuttgart/Ostfildern (Germany)

[31] Michaelis K, Höhn B-R, Oster P. Influence of lubricant on gear failures—Test methods and application to gearboxes in practice. Tribotest Journal. 2004;**11**(1):43-56

[32] Winter H, Richter M. Scuffing load capacity of hypoid and bevel gears: Conference proceedings of 8 JSME International Symposium "Gearing"; 1975

[33] Conrado E, Höhn B-R, Michaelis K, Klein M. Influence of oil supply on the scuffing load-carrying capacity of hypoid gears. Engineering Tribology. 2007;**221**:851-858

[34] Hadschuh RF. Testing of face-milled spiral bevel gears at high-speed and load. NASA/TM-2001-210743. 2001

[35] Akimov VV. Initial contour optimization for bevel gears with circular teeth. Journal of Machinery Manufacture and Reliability. 2008;**37**(4):371-378

[36] Suh S-H, Jung D-H, Lee E-S, Lee S-W. Modelling, implementation, and

manufacturing of spiral bevel gears with crown. International Journal of Advanced Manufacturing Technology. 2003;**21**:775-786

[37] Sung LM, Tsai YC. A study on the mathematical models and contact ratios of extended cycloid and cycloid bevel gear sets. Mechanism and Machine Theory. 1997;**32**:39-50

[38] Tsai YC, Hsu WY. The study on the design of spiral bevel gear sets with circular-arc contact paths and tooth profiles. Mechanism and Machine Theory. 2008;**43**:1158-1174

[39] Sekercioglu T, Kovan V. Pitting failure of truck spiral bevel gear. Engineering Failure Analysis. 2007;**14**:614-619

[40] Park M. Failure analysis of an accessory bevel gear installed on a J69 turbojet engine. Engineering Failure Analysis. 2003;**10**:371-382

[41] Fernandes PJL. Tooth bending fatigue failures in gears. Engineering Failure Analysis. 1996;**3**:219-225

[42] Michalczewski R, Kalbarczyk M, Tuszynski W, Szczerek M. The scuffing resistance of WC/C coated spiral bevel gears. Key Engineering Materials. 2014;**604**:36-40

[43] Tuszynski W, Kalbarczyk M, Michalak M, Michalczewski R, Wieczorek A. The effect of WC/C coating on the wear of bevel gears used in coal mines. Materials Science (Medžiagotyra). 2015;**21**(3):358-363

[44] Instruction Manual of the T-30 Back-to-Back Bevel Gear Test Rig. Radom: ITeE-PIB; 2014. (in Polish)

[45] Basiniuk UL, Levantsevich MA, Maksimchenko NN, Mardasevich AI. Improvement of triboengineering properties and noise reduction of tooth gears by cladding functional coatings on working surfaces of interfaced teeth. Journal of Friction and Wear. 2013;**34**(6):438-443

[46] Tomala AM, Michalczewski R, Osuch-Slomka E. Interaction of novel lubricant additives based on MoS_2 nanotubes with non-ferrous tribological materials. Tribologia. 2018;**280**(4):127-135

[47] Tomala A, Ripoll MR, Gabler C, Remškar M, Kalin M. Interactions between MoS_2 nanotubes and conventional additives in model oils. Tribology International. 2017;**110**:140-150

[48] Rodríguez Ripoll M, Tomala A, Gabler C, Dražić G, Pirker L, Remškar M. In situ tribochemical sulfurization of molybdenum oxide nanotubes. Nanoscale. 2018;**10**(7):3281-3290

[49] Tamura Y, Zhao H, Wang C, Morina A, Neville A. Interaction of DLC and B4C coatings with fully formulated oils in boundary lubrication conditions. Tribology International. 2016;**93**:666-680

[50] Kalin M, Oblak E, Akbari S. Evolution of the nano-scale mechanical properties of tribofilms formed from low- and high-SAPS oils and ZDDP on DLC coatings and steel. Tribology International. 2016;**96**:43-56

[51] Mandrino D, Podgornik B. XPS investigations of tribofilms formed on CrN coatings. Applied Surface Science. 2017;**396**:554-559

[52] Milewski K, Kudlinski J, Madej M, Ozimina D. The interaction between diamond like carbon (DLC) coatings and ionic liquids under boundary lubrication conditions. Meta. 2017;**56**:55-58

[53] Podgornik B, Zajec B, Strnad S, Stana-Kleinschek K. Influence of surface energy on the interactions between

hard coatings and lubricants. Wear. 2007;**262**:1199-1204

[54] Hassan MBH, Varman M, Mufti RA, Kalam MA, Zulkifli NWBM, Gulzar M. A review on effects of lubricant formulations on tribological performance and boundary lubrication mechanisms of non-doped DLC/DLC contacts. Critical Reviews in Solid State and Materials Sciences. 2017;**42**(4):267-294

[55] Shankar S, Krishnakumar P. Frictional characteristics of PVD coated mechanical seals against carbon under various classes of liquid lubricants. Industrial Lubrication Tribology. 2016;**68**(5):597-602

[56] Migranov MS, Migranov AM, Minigaleev SM, Shehtman SR. Tribological properties of multilayer coatings for cutting tool. Journal of Friction and Wear. 2018;**39**(3):245-250

[57] Wang B, Li J, Ma F, Xue Q. Tribological dependence of the high-performance ferrous-based coating on different coating counterparts in engine oil. Tribology Transactions. 2016;**59**(3):399-407

[58] Mannan A, Sabri MFM, Kalam MA, Hassan MH. Tribological performance of DLC/DLC and steel/DLC contacts in the presence of additivated oil. International Journal of Surface Science and Engineering. 2018;**12**(1):60-75